Springer Proceedings in Mathematics & Statistics

Volume 260

Springer Proceedings in Mathematics & Statistics

This book series features volumes composed of selected contributions from workshops and conferences in all areas of current research in mathematics and statistics, including operation research and optimization. In addition to an overall evaluation of the interest, scientific quality, and timeliness of each proposal at the hands of the publisher, individual contributions are all refereed to the high quality standards of leading journals in the field. Thus, this series provides the research community with well-edited, authoritative reports on developments in the most exciting areas of mathematical and statistical research today.

More information about this series at http://www.springer.com/series/10533

Paula Cerejeiras · Craig A. Nolder
John Ryan · Carmen Judith Vanegas Espinoza
Editors

Clifford Analysis and Related Topics

In Honor of Paul A. M. Dirac, CART 2014, Tallahassee, Florida, December 15–17

 Springer

Editors
Paula Cerejeiras
Departamento de Matemática
Universidade de Aveiro
Aveiro, Portugal

John Ryan
Department of Mathematics
University of Arkansas
Fayetteville, AR, USA

Craig A. Nolder
Department of Mathematics
Florida State University
Tallahassee, FL, USA

Carmen Judith Vanegas Espinoza
Department of Mathematics and
  Statistics, ICB
Technical University of Manabí
Portoviejo, Ecuador

ISSN 2194-1009   ISSN 2194-1017   (electronic)
Springer Proceedings in Mathematics & Statistics
ISBN 978-3-030-13080-0   ISBN 978-3-030-00049-3   (eBook)
https://doi.org/10.1007/978-3-030-00049-3

Mathematics Subject Classification (2010): 30G35, 15A66, 22E46, 53C27, 76W05

Preface

The present volume arises from the international conference *Clifford Analysis and Related Topics* held at the Florida State University, Tallahassee on December 2014. The conference was organized by Craig Nolder (Florida State University) and John Ryan (University of Arkansas) with the intent of celebrating the English theoretical physicist Paul Adrien Maurice Dirac, who died in Tallahassee in 1984, after spending his last decade of his life at Florida State University. P.A.M. Dirac made fundamental discoveries in the early formation of quantum mechanics. He shared the 1933 Nobel Prize in Physics with Erwin Schrödinger. He is the founder of the field of quantum electrodynamics. Notably, he developed a factorization of the Klein–Gordon equation which leads to the system of first-order Dirac equations which provided a relativistic wave equation for the electron. These equations provided a way to describe intrinsic spin and suggested the existence of antimatter, at first the positron which was discovered soon after the equations appeared. The equations turned out to describe all spin 1/2 particles, the fermions. The Dirac equations are based on a matrix representation of a Clifford algebra, now called Pauli matrices. Clifford algebras have found many applications in physics since this time including a role in the algebraic theory of the standard model of particle physics. Dirac was the Lucasian Professor of Mathematics at Cambridge from 1932 until 1969. He then came to Florida, working at Miami University, Coral Gables, and Florida State University, Tallahassee. He was a Visiting Professor at FSU during 1970–71 and accepted a Full Professorship in 1972. Dirac passed on August 8, 1984 and is buried in Roselawn Cemetery, Tallahassee, Fl.

Paul Dirac's work is at the very heart of Clifford Analysis, an active branch of mathematics that has grown significantly over the last 40 years and which covers both theoretical and applied physics. The field of Clifford Analysis began as a function theory for the solutions of the Dirac equation for spinor fields and, in such, can be regarded as a natural generalization to higher dimensions of the function theory of complex holomorphic functions.

The conference involved participants from Venezuela, Portugal, Brazil, Cape Verde, and USA, and this volume reflects not only the main contributions but also the stimulating and friendly atmosphere prevailing among the attendants.

Furthermore, the editors would like to express their gratitude to the anonymous referees without which this volume would have never seen the light.

We conclude with a statement of Dirac, published on *Scientific American*, May 1963:

> It seems to be one of the fundamental features of nature that fundamental physical laws are described in terms of a mathematical theory of great beauty and power, needing quite a high standard of mathematics for one to understand it.

It is our hope that the contributions on this volume make due honors to this statement.

Aveiro, Portugal   Paula Cerejeiras
Tallahassee, FL, USA   Craig A. Nolder
Fayetteville, AR, USA   John Ryan
Portoviejo, Ecuador   Carmen Judith Vanegas Espinoza
July 2018

The original version of the book frontmatter was revised: The fourth editor's affiliation has been corrected. The correction to the book frontmatter is available at https://doi.org/10.1007/978-3-030-00049-3_9

Contents

Contents

Lambda-Harmonic Functions: An Expository Account

K. Ballenger-Fazzone and C. A. Nolder

Abstract In this paper, we compile a variety of results on the $\lambda-$Laplacian operator, denoted by Δ_λ, a generalization of the well-known Laplacian in $\mathbb{R}^n$. We have compiled a list of known properties for Δ_λ when $\lambda = \frac{n-2}{2}$ and present analogous properties for Δ_λ. We close by discussing the $\lambda-$Poisson kernel, the function that solves the Dirichlet problem on the closed ball in $\mathbb{R}^n$.

Keywords Clifford analysis · Dirichlet problem · Lambda-Harmonic Laplacian · Poisson kernel

1 Introduction

The purpose of this paper is to compile a variety of interesting results on the $\lambda-$Laplacian operator, a generalization of the Laplacian Δ in $\mathbb{R}^n$. In Sect. 2, we define what it means for a function to be $\lambda-$harmonic and discuss how this operator is related to the Laplacian. We look at the special case when $\lambda = \frac{n-2}{2}$, known as the Invariant Laplacian, in Sect. 3 and present some properties of this operator. Section 4 provides a great deal of set up to show how some properties of the Invariant Laplacian do not generalize when $\lambda \neq \frac{n-2}{2}$. Finally, in Sect. 5, we discuss $\lambda-$Poisson kernel, which turns out to be the solution to the Dirichlet problem for $\lambda-$harmonic functions on the unit ball, and prove some new results for this kernel. Section 5 helps us to set up our next paper where we will solve the Dirichlet problem for $\lambda-$harmonic functions on an annular domain.

K. Ballenger-Fazzone (✉) · C. A. Nolder
Department of Mathematics, The Florida State University,
208 Love Building, 1017 Academic Way, Tallahassee, FL 32306-4510, USA
e-mail: kerr.fazzone@gmail.com

C. A. Nolder
e-mail: nolder@math.fsu.edu

© Springer Nature Switzerland AG 2018
P. Cerejeiras et al. (eds.), *Clifford Analysis and Related Topics*,
Springer Proceedings in Mathematics & Statistics 260,
https://doi.org/10.1007/978-3-030-00049-3_1

1

2 λ-Harmonic Functions

We denote by $\mathbb{B}^n$ the unit ball of $\mathbb{R}^n$ centered at the origin, where $n \geq 2$. A point $x \in \mathbb{B}^n$ is denoted $x = (x_1, x_2, \ldots, x_n)$. When the center or radius is important to us, we will denote by $B(a, r)$ an open ball centered at $a \in \mathbb{R}^n$ with radius $r > 0$. We denote the boundaries by S^{n-1} and $S(a, r)$ respectively.

Definition 1 Let $\lambda \in \mathbb{R}$. A function $u \in C^2(\mathbb{B}^n)$ is λ-**harmonic** if

$$\Delta_\lambda u = 0$$

in $\mathbb{B}^n$, where

$$\Delta_\lambda \stackrel{\text{DEF}}{=} \left(1 - |x|^2\right) \left[\frac{1 - |x|^2}{4} \Delta + \lambda E + \lambda \left(\frac{n-2}{2} - \lambda\right)\right], \tag{1}$$

is the λ-Laplacian,

$$\Delta = \sum_{i=1}^{n} \frac{\partial^2}{\partial x_i^2},$$

is the Laplacian on $\mathbb{R}^n$, and

$$E = \sum_{i=1}^{n} x_i \frac{\partial}{\partial x_i}$$

is the Euler operator.

Therefore,

$$\Delta_\lambda u = \left(1 - |x|^2\right) \left[\frac{1 - |x|^2}{4} \Delta u + \lambda E u + \lambda \left(\frac{n-2}{2} - \lambda\right) u\right].$$

If $\Delta_\lambda u = 0$, then u is an eigenvector of the differential operator

$$\frac{1 - |x|^2}{4} \Delta + \lambda E.$$

That is,

$$\frac{1 - |x|^2}{4} \Delta u + \lambda E u = \lambda \left(\frac{2-n}{2} + \lambda\right) u.$$

The λ-Laplacian is a generalization of two well-known operators:

1. If $\lambda = 0$, then

$$\Delta_0 = \frac{(1 - |x|^2)^2}{4} \Delta.$$

Thus solutions to $\Delta_0 u = 0$ are called harmonic.

2. If $\lambda = \dfrac{n-2}{2}$, then

$$\Delta_{\frac{n-2}{2}} = \frac{1}{4}\tilde{\Delta},$$

where

$$\tilde{\Delta} = (1 - |x|^2)^2 \Delta + 2(n-2)(1 - |x|^2)E$$

is the invariant Laplacian (or Laplace-Beltrami operator with respect to the Poincaré metric on $\mathbb{B}^n$). We call solutions to $\tilde{\Delta} u = 0$ **invariant harmonic** (or $\mathcal{M}$-harmonic) [18].

Before we discuss more about the λ-Laplacian, we first look at the invariant Laplacian in a different light.

3   The Invariant Laplacian $\left(\lambda = \frac{n-2}{2}\right)$

The invariant Laplacian $\tilde{\Delta}$ is the Laplace-Beltrami operator with respect to the Poincaré metric $ds = \dfrac{2|dx|}{1 - |x|^2}$ on $\mathbb{B}^n$. We can also define $\tilde{\Delta}$ in a geometric way. We remark that the content from this subsection comes from [18].

Definition 2 Let Ω be an open subset of $\mathbb{B}^n$ with $f \in C^2(\mathbb{B}^n)$ and $a \in \mathbb{B}^n$. We define the **invariant Laplacian** by

$$(\tilde{\Delta} f)(a) = \Delta(f \circ \phi_a)(0),$$

where

$$x^* = \begin{cases} x/|x|^2, & \text{if } x \neq 0 \\ 0, & \text{if } x = \infty \\ \infty, & \text{if } x = 0, \end{cases}$$

$$\phi_a(x) = \psi_a(x)^* = \frac{\psi_a(x)}{|\psi_a(x)|^2},$$

and

$$\psi_a(x) = a + (1 - |a|^2)(a - x)^*.$$

It is easy to see that ψ_a is a Möbius transformation mapping 0 to a^* and a to ∞.

Definition 3 Let Ω be an open subset of $\mathbb{B}^n$ with $f \in C^1(\mathbb{B}^n)$ and $a \in \mathbb{B}^n$. We define the **invariant gradient** by

$$(\tilde{\nabla} f)(a) = -\nabla(f \circ \phi_a)(0),$$

where $\nabla = \left(\frac{\partial}{\partial x_1}, \ldots, \frac{\partial}{\partial x_n} \right)$ is the usual gradient.

Remark 1 The minus sign in the definition above ensures that both ∇u and $\tilde{\nabla} u$ point in the same direction.

Let $f \in C^2(\mathbb{B}^n)$ and let $y = \psi(x)$ be a C^2 map from $\mathbb{B}^n$ into $\mathbb{B}^n$. If $g = f \circ \psi$, then

$$\nabla g(x) = \psi'(x) \nabla f(\psi(x))$$

and

$$\Delta g(x) = \sum_{i,j=1}^{n} \frac{\partial^2 f}{\partial y_i \partial y_j} \langle \nabla y_i, \nabla y_j \rangle + \sum_{j=1}^{n} \frac{\partial f}{\partial y_i} \Delta y_i,$$

where $\psi'(x)$ is the Jacobian matrix of ψ and $\langle \cdot, \cdot \rangle$ is the standard inner product in $\mathbb{R}^n$. Setting $y = \phi_a(x)$, we have that

$$\tilde{\nabla} f(a) = (1 - |a|^2) \nabla f(a)$$

and

$$\tilde{\Delta} f(a) = (1 - |a|^2)^2 \Delta f(a) + 2(n-2)(1 - |a|^2)\langle a, \nabla f(a) \rangle$$

as before.

Solutions to the invariant Laplacian are invariant under Möbius transformations of $\mathbb{B}^n$. We denote the group of Möbius transformations that leave $\mathbb{B}^n$ invariant by $\mathcal{M}(\mathbb{B}^n)$.

Theorem 1 *Let $f \in C^2(\mathbb{B}^n)$ and $\psi \in \mathcal{M}(\mathbb{B}^n)$. Then*

$$\tilde{\Delta}(f \circ \psi) = (\tilde{\Delta} f) \circ \psi$$

and

$$|\tilde{\nabla}(f \circ \psi)| = |(\tilde{\nabla} f) \circ \psi|.$$

Proof The proof of the first equality can be found in [18]. For the second, we mimic the proof from [15]. Let $b \in \psi^{-1}(\Omega)$ and set $a = \psi(b)$. Then we see that $\phi_a \circ \psi \circ \phi_b$ is a Möbius transformation of $\mathbb{B}^n$ that fixes 0. Therefore $\psi \circ \phi_b = \phi_a \circ A$, where A is some orthogonal transformation. Thus

$$|\tilde{\nabla}(f \circ \psi)(b)| = |-\nabla(f \circ \psi \circ \phi_b)(0)|$$
$$= |-\nabla(f \circ \phi_a \circ A)(0)|$$
$$= |-\nabla(f \circ \phi_a)(0)|$$
$$= |\tilde{\nabla}(f)(a)|$$
$$= |\tilde{\nabla}(f)(\psi(b))|.$$

We now present a few nice facts about the invariant Laplacian.

Theorem 2 *Invariance Properties of $\tilde{\Delta}$*

1. $\tilde{\Delta}$ *is a linear operator mapping* $C^2(\mathbb{B}^n) \to C(\mathbb{B}^n)$
2. *translations of invariant harmonic functions are invariant harmonic*
3. $r-$*dilates of invariant harmonic functions are invariant harmonic*
4. $\tilde{\Delta}$ *commutes with orthogonal transformations*

Proof To prove the first property, it is enough to assume that $u, v \in C^2(\mathbb{B}^n)$ and $c, d \in \mathbb{R}$. Then

$$\tilde{\Delta}(cu + dv) = \left(1 - |x|^2\right)\left[\frac{1 - |x|^2}{4}\Delta(cu + dv) + \left(\frac{n-2}{2}\right)E(cu + dv)\right]$$
$$= \left(1 - |x|^2\right)\left[\frac{1 - |x|^2}{4}(\Delta(cu) + \Delta(dv)) + \left(\frac{n-2}{2}\right)E(cu) + \left(\frac{n-2}{2}\right)E(dv)\right]$$
$$= c\tilde{\Delta}(u) + d\tilde{\Delta}(v).$$

The proof of 2. is clear.

To prove 3, we first assume $r \in \mathbb{R}$ with $r > 0$ and define $u_r(x) = u(rx)$, for $x \in (1/r)\mathbb{B}^n$. Then direct calculations show that

$$\Delta(u_r) = r^2(\Delta u)_r$$

and

$$E(u_r) = r^2(Eu)_r.$$

It follows that, if $\tilde{\Delta}u = 0$, then

$$\tilde{\Delta}(u_r) = \left(1 - |x|^2\right)\left[\frac{1 - |x|^2}{4}\Delta(u_r) + \left(\frac{n-2}{2}\right)E(u_r)\right]$$
$$= \left(1 - |x|^2\right)\left[r^2\frac{1 - |x|^2}{4}(\Delta u)_r + r^2\left(\frac{n-2}{2}\right)(Eu)_r\right]$$
$$= r^2(\tilde{\Delta}u)_r$$
$$= 0.$$

To prove *4*, we must show that if T is an orthogonal transformation and $u \in C^2(\mathbb{B}^n)$, then

$$\tilde{\Delta}(u \circ T) = (\tilde{\Delta}u) \circ T$$

on $T^{-1}(\mathbb{B}^n)$. We proceed following an argument from [2]. Let $[t_{jk}]$ denote the matrix for T relative to the standard basis in $\mathbb{R}^n$. Then

$$\frac{\partial}{\partial x_m}(u \circ T) = \sum_{j=1}^{n} t_{jm} \left(\frac{\partial}{\partial x_j} u \right) \circ T. \tag{2}$$

It is easy to see from (2) that

$$E(u \circ T) = (Eu) \circ T$$

and differentiating (2) shows

$$\Delta(u \circ T) = (\Delta u) \circ T.$$

Putting these two together then shows that

$$\tilde{\Delta}(u \circ T) = (\tilde{\Delta}u) \circ T.$$

We conclude this survey of the invariant Laplacian by listing the invariant analogous of some classical results from harmonic analysis.

Definition 4 Let Ω be an open subset of $\mathbb{B}^n$. A function $f \in C^2(\mathbb{B}^n)$ is **invariant subharmonic** (or $\mathscr{M}$-subharmonic) on Ω if $\tilde{\Delta}f(x) \geq 0$ for all $x \in \Omega$.

Remark 2 It is easy to prove that if f is invariant harmonic (invariant subharmonic) on $\mathbb{B}^n$, then $f \circ \psi$ is invariant harmonic (invariant subharmonic) on $\mathbb{B}^n$, for all $\psi \in \mathscr{M}(\mathbb{B}^n)$.

We can extend a mean-value property to invariant subharmonic functions.

Theorem 3 *Invariant Subharmonic Mean-Value Property*
Let Ω be an open subset of $\mathbb{B}^n$ and let $f \in C^2(\Omega)$. Then f is invariant subharmonic on Ω if and only if for all $a \in \Omega$

$$f(a) \leq \int_{\mathbb{S}^{n-1}} f(\phi_a(rt))d\sigma(t)$$

for all $r > 0$ such that $E(a, r) \subset \Omega$, where σ denotes the normalized surface measure on $\mathbb{S}^{n-1}$ and $E(a, r)$ is the Euclidean ball centered at a with radius r. f is invariant harmonic on Ω if and only if the equality holds.

The proof can be found in [18]. Invariant harmonic functions also satisfy a maximum principle.

Theorem 4 *Invariant Subharmonic Harmonic Maximum Principle*
Let Ω be an open subset of $\mathbb{B}^n$ and let $f \in C^2(\Omega)$ such that f is invariant subharmonic in Ω and continuous on $\overline{\Omega}$. If $f \leq 0$ on $\partial \Omega$, then $f \leq 0$ in Ω.

The proof can be found in [18]. More information on the invariant Laplacian on $\mathbb{B}^n$ can be found in [6, 9], whereas [15] discusses the invariant Laplacian on the unit ball in $\mathbb{C}^n$.

4 λ-Harmonic Functions Continued

In order to continue our discussion on Δ_λ, we must first review some preliminaries: the hypergeometric function and spherical harmonics.

4.1 The Hypergeometric Function

We define the (rising) Pochhammer symbol $(a)_l$ for an arbitrary $a \in \mathbb{C}$ and $l = 0, 1, \ldots$, by

$$(a)_l = \begin{cases} 1, & \text{if } l = 0 \\ a(a+1)\cdots(a+l-1), & \text{if } l > 0 \end{cases}$$

If a is not a negative integer, then

$$(a)_l = \frac{\Gamma(a+l)}{\Gamma(a)},$$

where Γ is the Gamma function defined on $\mathbb{C} \setminus \{-1, -2, \ldots\}$. Thus, for $x \in \mathbb{B}^n$, the **hypergeometric function** is defined to be

$$_2F_1(a, b; c; x) \stackrel{\text{DEF}}{=} \sum_{l=0}^{\infty} \frac{(a)_l (b)_l}{(c)_l} \frac{x^l}{l!},$$

and the series converges absolutely for all $x \in \mathbb{B}^n$ if $c - a - b > 0$. The function is undefined if c is a non-positive integer.

For convenience, we define

$$F_{\lambda,k}(x) \stackrel{\text{DEF}}{=} {}_2F_1\left(-\lambda, k + \frac{n-2}{2} - \lambda; k + \frac{n}{2}; x\right).$$

We remark that many of the proofs involving the hypergeometric function rely on various formulas found in [3, 11].

4.2 Homogeneous Harmonic Polynomials and Spherical Harmonics

We begin this subsection by discussing some classical results concerning homogeneous polynomials from harmonic analysis. We direct the reader to [2, 7, 12, 13] for more information.

Definition 5 A polynomial Y_m is **homogeneous of degree** m if Y_m is of the form

$$Y_m(x) \overset{\text{DEF}}{=\joinrel=} \sum_{|\alpha|=m} c_\alpha x^\alpha,$$

where

Alternatively, Y_m is homogeneous of degree m if, for all $t \in \mathbb{R}$,

$$Y_m(tx) = t^m Y_m(x).$$

It is well-known that every degree m polynomial Y on $\mathbb{R}^n$ can be written uniquely as

$$Y(x) = \sum_{j=0}^{m} Y_j(x),$$

where Y_j is homogeneous of degree j. It is then easy to see that Y is harmonic if and only if Y_j is harmonic for each $j = 0, 1, \ldots, m$.

Notation 5 *We denote by $\mathscr{P}_m(\mathbb{R}^n)$ the set of all homogeneous polynomials on $\mathbb{R}^n$ of degree m and by $\mathscr{H}_m(\mathbb{R}^n)$ the set of all homogeneous harmonic polynomials on $\mathbb{R}^n$ of degree m.*

We are able to decompose $\mathscr{P}_m(\mathbb{R}^n)$ into the direct sum of two subspaces, which we present in the following theorem [2].

Theorem 6 *If $m \geq 2$, then we can write*

$$\mathscr{P}_m(\mathbb{R}^n) \equiv \mathscr{H}_m(\mathbb{R}^n) \oplus |x|^2 \mathscr{P}_{m-2}(\mathbb{R}^n).$$

The proof of Theorem 6 relies on the fact that no multiple of the polynomial $|x|^2$ is harmonic (see Corollary 5.3 [2]).

To proceed further, we must introduce hyperspherical coordinates on n-dimensional Euclidean space. Our coordinate system consists of one radial coordinate r and $n - 1$ angular coordinates denoted by $\phi_1, \phi_2, \ldots, \phi_{n-1}$, where $\phi_{n-1} \in [0, 2\pi]$ and $\phi_i \in [0, \pi]$, for $i = 1, 2, \ldots, n - 2$. Then the relationship between Euclidean coordinates $x_1, \ldots x_n$ and hyperspherical coordinates is given by

$$x_1 = r\cos(\phi_1)$$
$$x_2 = r\sin(\phi_1)\cos(\phi_2)$$
$$x_3 = r\sin(\phi_1)\sin(\phi_2)\cos(\phi_3)$$
$$\vdots$$
$$x_{n-1} = r\sin(\phi_1)\sin(\phi_2)\cdots\sin(\phi_{n-2})\cos(\phi_{n-1})$$
$$x_n = r\sin(\phi_1)\sin(\phi_2)\cdots\sin(\phi_{n-1})$$

It follows that each point $x \in \mathbb{R}^n \setminus \{0\}$ can be written uniquely as $x = r\omega$, where $r = |x|$ and $|\omega| = 1$. Hence, we can write every $Y_m \in \mathscr{H}_m(\mathbb{R}^n)$ as

$$Y_m(x) = r^m Y_m(\omega). \tag{3}$$

The restriction to $\mathbb{S}^{n-1}$ yields a spherical harmonic of degree m. We make this clear in the following definition.

Definition 6 Let $Y_m \in \mathscr{H}_m(\mathbb{R}^n)$. The restriction of Y_m to $\mathbb{S}^{n-1}$ is called a **spherical harmonic of degree** m and is denoted by

$$\mathscr{H}_m(\mathbb{S}^{n-1}) \stackrel{\text{DEF}}{=\!=} \left\{ Y \upharpoonright_{\mathbb{S}^{n-1}} \mid Y \in \mathscr{H}_m(\mathbb{R}^n) \right\}.$$

We can see from (3) that every $Y_m \in \mathscr{H}_m(\mathbb{R}^n)$ is completely determined by its restriction to $\mathbb{S}^{n-1}$. Moreover, we remark that if $X_m, Y_m \in \mathscr{H}_m(\mathbb{R}^n)$ and $X_m = Y_m$ on $\mathbb{S}^{n-1}$, then $X_m = Y_m$ on all of $\mathbb{R}^n$ [12].

The interested reader can find more information on applications of spherical harmonics, including interesting applications in computer graphics, in [4, 5, 8]. Though not necessary for what follows, we include the following two theorems for completeness [2, 12, 13].

Theorem 7 $\mathscr{H}_m(\mathbb{R}^n)$ *is finite-dimensional, for all* $m = 0, 1, \ldots$. *Specifically,* $\dim \mathscr{H}_0(\mathbb{R}^n) = 1$, $\dim \mathscr{H}_1(\mathbb{R}^n) = n$, *and, if* $m \geq 2$, *then*

$$\dim \mathscr{H}_m(\mathbb{R}^n) = \binom{n+m-1}{n-1} - \binom{n+m-3}{n-1}.$$

Theorem 8 *If* $k \neq l$, *then the spaces* $\mathscr{H}_k(\mathbb{S}^{n-1})$ *and* $\mathscr{H}_l(\mathbb{S}^{n-1})$ *are orthogonal in* $L^2(\mathbb{S}^{n-1})$ *with respect to the inner product*

$$\langle f, g \rangle_{L^2(\mathbb{S}^{n-1})} = \int_{\mathbb{S}^{n-1}} f\overline{g}\,d\sigma. \tag{4}$$

Additionally,

$$L^2(\mathbb{S}^{n-1}) = \bigoplus_{k=0}^{n} \mathscr{H}_k(\mathbb{S}^{n-1}).$$

We see from (4) that $L^2(\mathbb{S}^{n-1})$ is the classical Hilbert space of all Borel-measurable functions on the sphere with the aforementioned inner product [2]. This brings us to a special class of spherical harmonics that are rotationally invariant.

Let $x \in \mathbb{S}^{n-1}$ and define the map

$$\Lambda : \mathcal{H}_m(\mathbb{S}^{n-1}) \to \mathbb{C}$$

by

$$\Lambda(Y) = Y(x).$$

We know that since $\mathcal{H}_m(\mathbb{S}^{n-1})$ is a finite-dimensional inner product space with respect to (4), the map Λ is bounded, and hence there exists a unique polynomial $Z_m(x, \cdot) \in \mathcal{H}_m(\mathbb{S}^{n-1})$ such that

$$Y(x) = \langle Y, Z_m(x, \cdot) \rangle = \int_{\mathbb{S}^{n-1}} Y(\zeta) \overline{Z_m(x, \zeta)} d\sigma(\zeta),$$

for all $Y \in \mathcal{H}_m(\mathbb{S}^{n-1})$ [2]. The spherical harmonic $Z_m(x, \zeta)$ is called the **zonal harmonic of degree m with pole ζ**. We remark that when $m = 0$, the zonal harmonics reduce to the associated Legendre polynomials [1, 8, 16].

We can rewrite the decomposition of $L^2(\mathbb{S}^{n-1})$ in Theorem 8 using zonal harmonics.

Theorem 9 *Let $f \in L^2(\mathbb{S}^{n-1})$ and let $Y_m(x) = \langle f, Z_m(x, \cdot) \rangle$ for $m \geq 0$ and $x \in \mathbb{S}^{n-1}$. Then $Y_m \in \mathcal{H}_m(\mathbb{S}^{n-1})$ and*

$$f = \sum_{m=0}^{\infty} Y_m$$

in $L^2(\mathbb{S}^{n-1})$.

The proof can be found in [2].

For the remainder of this chapter, we will relax our notation for spherical and zonal harmonics. Every element $Y \in \mathcal{H}_m(\mathbb{S}^{n-1})$ has a unique extension to $\mathcal{H}_m(\mathbb{R}^n)$. For this reason, we will use Y to denote this extension as well. Similarly, $Z_m(x, \cdot)$ will refer to the extension to $\mathcal{H}_m(\mathbb{R}^n)$.

One of the reasons that zonal harmonics are so useful is the following reproducing property. Let $x \in \mathbb{R}^n \setminus \{0\}$ and $Y \in \mathcal{H}_m(\mathbb{R}^n)$. Then

$$Y(x) = |x|^m Y\left(\frac{x}{|x|}\right)$$

$$= |x|^m \int_{\mathbb{S}^{n-1}} Y(\zeta) Z_m\left(\frac{x}{|x|}, \zeta\right) d\sigma(\zeta)$$

$$= \int_{\mathbb{S}^{n-1}} Y(\zeta) Z_m(x, \zeta) d\sigma(\zeta)$$

We direct the reader to [2, 17] for more information on zonal harmonics. For instance, Axler shows that there is an expansion of the Poisson kernel for $\mathbb{B}^n$ with zonal harmonics as coefficients. We will use zonal harmonics in our formula for the Poisson kernel on the annulus near the end of this chapter.

4.3 Linearity, Möbius Invariance, and Polyharmonicity of Δ_λ

Unfortunately, only the first statement in Theorem 2 has a Δ_λ analogue for $\lambda \neq \frac{n-2}{2}$. The other statements fail to be true due to the nonzero $\lambda(\frac{n-2}{2} - \lambda)$ term.

Theorem 10 *The λ-Laplacian is a linear operator mapping $C^2(\mathbb{B}^n) \to C(\mathbb{B}^n)$.*

Proof We follow the proof of Theorem 2. Assume that $u, v \in C^2(\mathbb{B}^n)$ and $c, d \in \mathbb{R}$. Then

$$
\Delta_\lambda(cu + dv) = \left(1 - |x|^2\right)\left[\frac{1 - |x|^2}{4}\Delta(cu + dv) + \lambda E(cu + dv) + \lambda\left(\frac{n-2}{2} - \lambda\right)(cu + dv)\right]
$$

$$
= \left(1 - |x|^2\right)\left[\frac{1 - |x|^2}{4}(\Delta(cu) + \Delta(dv)) + (\lambda E(cu) + \lambda E(dv))\right.
$$

$$
\left. + \left(\lambda\left(\frac{n-2}{2} - \lambda\right)(cu) + \lambda\left(\frac{n-2}{2} - \lambda\right)(dv)\right)\right]
$$

$$
= c\Delta_\lambda(u) + d\Delta_\lambda(v)
$$

The λ-Laplacian is also invariant under Möbius transformations of $\mathbb{B}^n$ [10].

Theorem 11 *For all $u \in C^2(\mathbb{B}^n)$ and $\psi \in \mathcal{M}(\mathbb{B}^n)$,*

$$
\Delta_\lambda\left[(det\psi'(x))^{\frac{n-2-2\lambda}{2n}}u(\psi(x))\right] = (det\psi'(x))^{\frac{n-2-2\lambda}{2n}}(\Delta_\lambda u)(\psi(x)),
$$

where ψ' is the Jacobian matrix of $\psi \in \mathcal{M}(\mathbb{B}^n)$.

One of the important boundary regularity results from [10] involves polyharmonicity. We will state the result in the next section; however, we include a definition and a theorem that will illustrate the flavor of the result.

Definition 7 Let Ω be an open subset of $\mathbb{R}^n$ and $u \in C^{2k}(\Omega)$. Then u is **polyharmonic of finite degree** k in Ω if $\Delta^k u(x) = 0$ for all $x \in \Omega$, where Δ^k is defined inductively by $\Delta^k = \Delta\Delta^{k-1}$.

The following theorem tells is that, "... if λ is a nonnegative integer, any solution of $\Delta_\lambda u = 0$ is polyharmonic of degree $\lambda + 1$ in $\mathbb{B}^n$" [10].

Theorem 12 *Let Ω be an open subset of $\mathbb{B}^n$ and let $u \in C^2(\Omega)$. If $\Delta_\lambda u = 0$ in Ω, then $\Delta_{\lambda-1}(\Delta u) = 0$.*

We now present the harmonic expansion of λ-harmonic functions which uses the hypergeometric function.

Theorem 13 *If u is λ-harmonic in $\mathbb{B}^n$, then*

$$u(x) = \sum_{k=0}^{\infty} {}_2F_1\left(-\lambda, k + \frac{n-2}{2} - \lambda; k + \frac{n}{2}; r^2\right) r^k u_k(\zeta),$$

where $x = r\zeta$, $|x| = r$, $|\zeta| = 1$, $u_k \in \mathcal{H}_k$, and where the series converges uniformly and absolutely on compact subsets in $\mathbb{B}^n$.

The proof of this theorem can be found in [10]. We remark that when $\lambda = 0$ we recover the usual harmonic expansion since ${}_2F_1(0, \cdot; \cdot; \cdot) = 1$.

5 The λ-Poisson Kernel on $\mathbb{B}^n$

Let $f \in C^2(\mathbb{S}^{n-1})$. As in the classical case, we are concerned with a function $u \in C^{\infty}(\overline{\mathbb{B}^n})$, with $f \not\equiv 0$, that solves the Dirichlet problem

$$\begin{cases} \Delta_\lambda u = 0 & \text{in } \mathbb{B}^n, \\ u = f & \text{on } \mathbb{S}^{n-1} \end{cases} \tag{5}$$

The function we are looking for is the λ-Poisson kernel, which we define next. We direct the reader to [10] for more information on this function.

Definition 8 The λ-Poisson kernel is the function

$$P_\lambda(x, \zeta) = c(\lambda, n) \frac{(1 - |x|^2)^{1+2\lambda}}{|x - \zeta|^{n+2\lambda}},$$

where

$$c(\lambda, n) = \frac{\Gamma(\frac{n}{2} + \lambda)\Gamma(1 + \lambda)}{\Gamma(\frac{n}{2})\Gamma(1 + 2\lambda)}.$$

The proof of the following theorem can be found in [10].

Theorem 14 *The λ-Poisson kernel is λ-harmonic. That is*

$$\Delta_\lambda P_\lambda = 0.$$

Furthermore, for all $f \in C^2(\mathbb{S}^{n-1})$, the function

$$P_\lambda[f](x) = \int_{\mathbb{S}^{n-1}} P_\lambda(x, \zeta) f(\zeta) d\sigma(\zeta)$$

solves the Dirichlet problem (5) for all $u \in C^2(\mathbb{S}^{n-1})$ if and only if $\lambda > -1/2$.

Now, if $f = \sum_k Y_k$ is the spherical expansion of f, then we can rewrite the λ-Poisson kernel using the hypergeometric function as

$$\begin{aligned}
u(x) &= \sum_{k=0}^{\infty} \frac{{}_2F_1\left(-\lambda, k + \frac{n-2}{2} - \lambda; k + \frac{n}{2}; r^2\right)}{{}_2F_1\left(-\lambda, k + \frac{n-2}{2} - \lambda; k + \frac{n}{2}; 1\right)} |x|^k Y_k(\zeta) \\
&= \frac{F_{\lambda,k}(r^2)}{F_{\lambda,k}(1)} |x|^k Y_k(\zeta).
\end{aligned}$$

We are now equipped to present the following theorem, the proof of which is found in [10].

Theorem 15 *Let $f \in C^\infty(\mathbb{S}^{n-1})$ with $f \not\equiv 0$. Then the solution u to the Dirichlet problem (5) is in $C^\infty(\overline{\mathbb{B}^n})$ if and only if one of the following occurs:*

1. λ is a nonnegative integer,
2. the data f has a finite spherical harmonic expansion

$$f = \sum_{k=0}^{N} Y_k, \quad Y_k \in \mathcal{H}_k(\mathbb{S}^{n-1}),$$

and $\lambda + \frac{2-n}{2} - N$ is a nonnegative integer, where N is the greatest index k such that $Y_k \not\equiv 0$ on $\mathbb{S}^{n-1}$.

The proof of this theorem shows the following corollary which involves polyharmonicity.

Corollary 1 *If $\lambda > -1/2$ is neither an integer nor a half-integer not less than $\frac{n-2}{2}$, then one cannot expect $C^{2+[2\lambda]}$ up to the boundary solutions of the Dirichlet problem (5). Moreover, if $u \in C^{2+[2\lambda]}(\overline{\mathbb{B}^n})$ is λ-harmonic, then u is either a polynomial of degree at most $2\lambda + 2 - n$ or a polyharmonic function of degree $\lambda + 1$.*

As before, we define a multi-index α to be an n-tuple $(\alpha_1, \ldots, \alpha_n)$, where α_j is a nonnegative integer for each $1 \leq j \leq n$. For a multi-index α, let D^α be the partial differential operator $D_1^{\alpha_1} \cdots D_n^{\alpha_n}$, where D_j^0 is the identity. Now follows our first result.

Theorem 16 *Suppose (u_m) is a sequence of λ-harmonic functions on Ω such that (u_m) converges uniformly to a function u on each compact subset of Ω. Then u is λ-harmonic on Ω. Moreover, for each multi-index α, $D^\alpha u_m$ converges uniformly to $D^\alpha u$ on each compact subset of Ω.*

Proof We mimic a proof from [2]. Let $B(a, r)$ be an open ball centered at a with radius r and $\overline{B}(a, r) \subset \Omega$. It suffices to show that u is λ-harmonic on $B(a, r)$.

Without loss of generality, assume that $B(a, r) = \mathbb{B}^n$. For every $x \in \mathbb{B}^n$ and every m, we have that

$$u_m(x) = \int_{S^{n-1}} P_\lambda(x, \zeta) u_m(\zeta) d\sigma(\zeta).$$

Since $u_m(x)$ converges uniformly to $u(x)$ on each compact subset of Ω, we can pass the limit inside the integral to get

$$\begin{aligned}
\lim_{m \to \infty} u_m(x) &= \lim_{m \to \infty} \int_{S^{n-1}} P_\lambda(x, \zeta) u_m(\zeta) d\sigma(\zeta) \\
&= \int_{S^{n-1}} P_\lambda(x, \zeta) \left(\lim_{m \to \infty} u_m(\zeta) \right) d\sigma(\zeta) \\
&= \int_{S^{n-1}} P_\lambda(x, \zeta) u(\zeta) d\sigma(\zeta) \\
&= u(x).
\end{aligned}$$

Hence, since P_λ is λ-harmonic, its partial derivatives exist and are continuous, we can pass the λ-Laplacian inside the integral to get

$$\begin{aligned}
\Delta_\lambda u(x) &= \Delta_\lambda \int_{S^{n-1}} P_\lambda(x, \zeta) u(\zeta) d\sigma(\zeta) \\
&= \int_{S^{n-1}} \Delta_\lambda P_\lambda(x, \zeta) u(\zeta) d\sigma(\zeta) \\
&= 0.
\end{aligned}$$

Thus u is λ-harmonic.

Now, let α be a multi-index and let $x \in \mathbb{B}^n$. Since u_m is harmonic for each m, we can pass the partial differential operator inside the integral to get

$$\begin{aligned}
D^\alpha u_m(x) &= D^\alpha \int_{S^{n-1}} P_\lambda(x, \zeta) u_m(\zeta) d\sigma(\zeta) \\
&= \int_{S^{n-1}} D^\alpha P_\lambda(x, \zeta) u_m(\zeta) d\sigma(\zeta) \\
&\longrightarrow \int_{S^{n-1}} D^\alpha P_\lambda(x, \zeta) u(\zeta) d\sigma(\zeta) \\
&= D^\alpha u(x).
\end{aligned}$$

If $K \subset \mathbb{B}^n$ is compact, then $D^\alpha P_\lambda$ is uniformly bounded on $K \times S$, and hence $D^\alpha u_m$ converges to $D^\alpha u$ uniformly on all of K.

The next two results are found in [14]. We can use the λ-Poisson kernel to prove a monotonicity property of λ-harmonic functions and characterize their behavior on rays.

Theorem 17 *Let u be a positive λ-harmonic function defined in $\mathbb{B}^n$ by a positive Borel measure μ on $\mathbb{S}^{n-1}$ with the Poisson kernel P_λ. For $\zeta \in \mathbb{S}^{n-1}$, if $\lambda > -\frac{n}{2}$ $(\lambda < -\frac{n}{2})$, the function*

$$\frac{(1-r)^{n-1}}{(1+r)^{1+2\lambda}} u(r\zeta)$$

is decreasing (increasing) for $0 \le r < 1$, and the function

$$\frac{(1+r)^{n-1}}{(1-r)^{1+2\lambda}} u(r\zeta)$$

is increasing (decreasing) for $0 \le r < 1$. Also,

$$\lim_{r \to 1}(1-r)^{n-1} u(r\zeta) = \begin{cases} 2^{1+2\lambda}\mu(\{\zeta\}), & \lambda > -n/2 \\ \infty, & \lambda < -n/2, \, \mu(\{\zeta\}^c) > 0 \\ 2^{1+2\lambda}\mu(\{\zeta\}), & \lambda < -n/2, \, \mu(\{\zeta\}^c) = 0 \end{cases}$$

and

$$\lim_{r \to 1} \frac{u(r\zeta)}{(1-r)^{1+2\lambda}} = \int_{\mathbb{S}^{n-1}} \frac{2^{1+2\lambda}}{|\zeta - \xi|^{n+2\lambda}} d\mu(\xi).$$

Theorem 18 *Let u be a positive λ-harmonic function defined in $\mathbb{B}^n$ by a positive Borel measure μ on $\mathbb{S}^{n-1}$ with the Poisson kernel P_λ. For $\zeta \in \mathbb{S}^{n-1}$ and $0 \le r' \le r < 1$.*

If $\lambda > -\frac{n}{2}$, then

$$\left(\frac{1-r}{1-r'}\right)^{2\lambda+1} \left(\frac{1+r'}{1+r}\right)^{n-1} u(r'\zeta) \le u(r\zeta) \le \left(\frac{1+r}{1+r'}\right)^{2\lambda+1} \left(\frac{1-r'}{1-r}\right)^{n-1} u(r'\zeta).$$

If $\lambda < -\frac{n}{2}$, then

$$\left(\frac{1+r}{1+r'}\right)^{2\lambda+1} \left(\frac{1-r'}{1-r}\right)^{n-1} u(r'\zeta) \le u(r\zeta) \le \left(\frac{1-r}{1-r'}\right)^{2\lambda+1} \left(\frac{1+r'}{1+r}\right)^{n-1} u(r'\zeta).$$

For $r' = 0$, the above becomes

$$\frac{(1-r)^{2\lambda+1}}{(1+r)^{n-1}} u(0) \le u(r\zeta) \le \frac{(1+r)^{2\lambda+1}}{(1-r)^{n-1}} u(0)$$

for $\lambda > -\frac{n}{2}$, *and*

$$\frac{(1+r)^{2\lambda+1}}{(1-r)^{n-1}}u(0) \leq u(r\zeta) \leq \frac{(1-r)^{2\lambda+1}}{(1+r)^{n-1}}u(0)$$

for $\lambda < -\frac{n}{2}$.

6   Conclusion

The details in Sect. 5 help us to solve a similar problem on annular domains, which we will present in our next paper. To do this, we must use the λ-Kelvin transform.

Definition 9 Let u be a function defined on a set $\Omega \subset \widehat{\mathbb{R}^n}$. We define the λ-Kelvin transform of u on Ω^* by

$$K_\lambda[u](x) = (-1)^{-(1+2\lambda)}|x|^{2\lambda+2-n}u(x^*).$$

We are then able to prove that, for $\Omega \subset \mathbb{R}^n \setminus \{0\}$, if u is λ-harmonic in Ω, then $K_\lambda[u]$ is λ-harmonic in Ω^*.

Putting all of this together, we use the $\lambda-$Kelvin transform to prove our main result. Let $r_0 \in \mathbb{R}$, with $0 < r_0 < 1$, and define the annulus $A = \{x \in \mathbb{R}^n \mid r_0 < |x| < 1\}$. For convenience, we recall that

$$F_{\lambda,k}(x) \overset{\text{DEF}}{=\!=} {}_2F_1\left(-\lambda, k + \frac{n-2}{2} - \lambda; k + \frac{n}{2}; x\right).$$

Theorem 19 *Poisson Kernel on Annulus*
Let f *be continuous on* ∂A. *Define* u *on* $\overline{A}$ *by*

$$u(x) = \begin{cases} P_\lambda^A[f](x), & if\, x \in A \\ f(x), & if\, x \in \partial A, \end{cases}$$

where

$$P_\lambda^A[f](x) = \int_{\mathbb{S}^{n-1}} P_\lambda^A(x, \zeta)f(\zeta)d\sigma(\zeta) + \int_{\mathbb{S}^{n-1}} P_\lambda^A(x, r_0\zeta)f(r_0\zeta)d\sigma(\zeta),$$

$$P_\lambda^A(x, \zeta) = \sum_{k=0}^{\infty} b_k(x)Z_k(x, \zeta),$$

$$P_\lambda^A(x, r_0\zeta) = \sum_{k=0}^{\infty} c_k(x) Z_k(x, \zeta),$$

$$b_k(x) = \frac{F_{\lambda,k}(\frac{1}{r_0}) r_0^{(2\lambda+2-n-2k)} F_{\lambda,k}(|x|) - F_{\lambda,k}(r_0) F_{\lambda,k}(\frac{1}{|x|}) |x|^{(2\lambda+2-n-2k)}}{F_{\lambda,k}(\frac{1}{r_0}) r_0^{(2\lambda+2-n-2k)} F_{\lambda,k}(1) - F_{\lambda,k}(r_0) F_{\lambda,k}(1)},$$

and

$$c_k(x) = \frac{F_{\lambda,k}(|x|) - F_{\lambda,k}(\frac{1}{|x|}) |x|^{(2\lambda+2-n-2k)}}{F_{\lambda,k}(|r_0|) - F_{\lambda,k}(\frac{1}{r_0}) |r_0|^{(2\lambda+2-n-2k)}}.$$

Then u is continuous on $\overline{A}$ and λ-harmonic on A.

References

1. Abramowitz, M., Stegun, I.A.: Handbook of Mathematical Functions: with Formulas, Graphs, and Mathematical Tables. Dover Books on Mathematics. Dover Publications (2012)
2. Axler, S., Bourdon, P., Ramey, W.: Harmonic Function Theory, 2 edn. Springer (1992)
3. Bateman, H., Erdélyi, A.: Higher Transcendental Functions. McGraw-Hill, New York (1953)
4. Bromwich, T.J.I.: Investigations on series of zonal harmonics. In: Proceedings of the London Mathematical Society, vol. s2-4(1), pp. 204–222 (1907)
5. Bezubik, A., Dabrowska, A., Strasburger, A.: On spherical expansions of zonal functions on Euclidean spheres. Archiv der Mathematik **90**(1), 70–81 (2008)
6. Cho, S., Choe, B.R., Koo, H.: Weak Hopf lemma for the invariant Laplacian and related elliptic operators. JMAA **408**, 576–588 (2013)
7. Dai, F., Xu, Y.: Approximation Theory and Harmonic Analysis on Spheres and Balls. Springer Monographs in Mathematics, Springer, New York (2013)
8. Groemer, H.: Geometric Applications of Fourier Series and Spherical Harmonics, vol. 61. Cambridge University Press, New York (1996)
9. Kwon, E.G.: One radius theorem for the eigenfunctions of the invariant Laplacian. PAMS **116**(1), 27–34 (1992)
10. Liu, C., Peng, L.: Boundary regularity in the Dirichlet problem for the invariant Laplacians Δ_γ on the unit real ball. PAMS **132**(1), 3259–3268 (2004)
11. Magnus, W., Oberhettinger, F., Soni, R.P.: Formulas and Theorems for the Special Functions of Mathematical Physics, vol. 52, 3rd edn. Springer, New York (1966)
12. Morais, J.: Approximation by homogeneous polynomial solutions of the Riesz system in $\mathbb{R}^3$. PhD thesis, Institute of Mathematics and Physics, Bauhaus University Weimar (2009)
13. Müller, C.: Spherical Harmonics. Lecture Notes in Mathematics, vol. 17. Springer-Verlag, Berlin (1966)
14. Pan Y., Wang, M.: On the monotonicity of positive invariant harmonic functions in the unit ball. arXiv:math/0702064v1 [math.CA]
15. Rudin, W.: Function Theory in the Unit Ball of $\mathbb{C}^n$. Classics in Mathematics. Springer (2008)
16. Schach, S.R.: New identities for legendre associated functions of integral order and degree I. SIAM J. Math. Anal. **7**(1), 59–69 (1976)
17. Stein, E.M., Weiss, G.L.: Introduction to Fourier Analysis on Euclidean Spaces. Princeton University Press, Princeton, NJ (1971)
18. Stoll, M.: Harmonic Function Theory on Real Hyperbolic Space, Lecture Notes, Preliminary Draft, http://www.math.sc.edu/math/sites/math.sc.edu/files/attachments/hyperbolic.pdf?q=math/sites/sc.edu.math/files/attachments/hyperbolic.pdf

Some Applications of Parabolic Dirac Operators to the Instationary Navier-Stokes Problem on Conformally Flat Cylinders and Tori in $\mathbb{R}^3$

P. Cerejeiras, U. Kähler and R. S. Kraußhar

Abstract In this paper we give a survey on how to apply recent techniques of Clifford analysis over conformally flat manifolds to deal with instationary flow problems on cylinders and tori. Solutions are represented in terms of integral operators involving explicit expressions for the Cauchy kernel that are associated to the parabolic Dirac operators acting on spinor sections of these manifolds.

Keywords Instationary PDE · Navier-Stokes equations · Conformally flat manifolds · Clifford algebra valued integral calculus

1 Introduction

The treatment of Navier-Stokes systems is a principal topic in mathematical physics, as they are a main model for describing physical phenomena linked to Newtonian flow from the water flow in a pipe to air flow around a wing. Thus, these systems form the basis of fluid dynamics. The literature addressing these kind of problems is large and abundant ranging from the development of efficient numerical methods up to theoretical analysis in the hope to gain some more insight into the structure and nature

P. Cerejeiras · U. Kähler
CIDMA, Departamento de Matemática, Universidade de Aveiro, Campus de Santiago, 3810-193 Aveiro, Portugal
e-mail: pceres@ua.pt

U. Kähler
e-mail: ukaehler@ua.pt

R. S. Kraußhar (✉)
Professur für Mathematik, Erziehungswissenschaftliche Fakultät,
Universität Erfurt, Nordhäuser Str. 63, D-99089 Erfurt, Germany
e-mail: soeren.krausshar@uni-erfurt.de

© Springer Nature Switzerland AG 2018   19
P. Cerejeiras et al. (eds.), *Clifford Analysis and Related Topics*,
Springer Proceedings in Mathematics & Statistics 260,
https://doi.org/10.1007/978-3-030-00049-3_2

of solutions in some special cases. The proof of existence of strong solutions over all times is still open and belongs to the millennium prize problems. This shows the importance of the field as well as the growing need for further research of particular theoretical aspects of said systems related to fluid dynamics.

In this paper we revisit the three dimensional instationary Navier-Stokes equations (NSE) for incompressible fluids

$$- \Delta \mathbf{u} + \frac{\partial \mathbf{u}}{\partial t} + (\mathbf{u} \, \text{grad}) \, \mathbf{u} + \text{grad} \, p = \mathbf{f}, \quad \text{in } G \tag{1}$$

$$\text{div } \mathbf{u} = 0, \quad \text{in } G \tag{2}$$

$$\mathbf{u} = \mathbf{0}, \quad \text{on } \partial G. \tag{3}$$

Here, $\mathbf{u}$ represents the velocity of the flow, p stands for the pressure and $\mathbf{f}$ for the specific body force.

Since the 1980s quaternionic analysis, and more generally Clifford analysis, proves to be a powerful tool for the study of this type of non-linear PDE system. In the Stokes problem, which can be regarded as a much simpler stationary and linear version of the NSE, the Laplacian $\Delta = \sum_{i=1}^{3} \frac{\partial^2}{\partial x_i^2}$ can be factorized by a linear first order operator, namely the Euclidean Dirac operator $D := \sum_{i=1}^{3} \frac{\partial}{\partial x_i} e_i$ viz $D^2 = -\Delta$. Analogously to the Cauchy-Riemann operator in two dimensions, the Dirac operator is the basis for a rich function theory in higher dimensions. One obtains Cauchy's integral formula in complete analogy with the standard complex Cauchy formula as the basis for the development of further interesting results, see for instance [10] and elsewhere. Similar as in the complex case the resulting function theory provides a refinement of classical harmonic analysis. Furthermore, since the Dirac operator acts on spinor-valued functions, it encodes much more geometrical information than the classical Laplace operator which acts on scalar-valued functions.

The quaternionic operator calculus for elliptic boundary value problems was developed in several works of K. Gürlebeck, W. Sprößig, M. Shapiro, V.V. Kravchenko and many others, see for instance [13, 14, 17, 18]. In these works a remarkable number of stationary linear and non-linear boundary value problems have been addressed successfully by means of singular integral operators. The quaternionic calculus actually leads to further new explicit criteria for regularity, existence and uniqueness of the solutions of these systems. Based on those new theoretical results also new numerical algorithms based on discrete version of the quaternionic calculus were developed, see for instance [11, 12], or the book of M. Mitrea [21] in which the study of singular integral operators and Clifford wavelets has successfully been applied to boundary value probems over Lipschitz domains, see also his follow-up work [22]. Also fully analytic representation formulas for the solutions for the Navier-Stokes equations and for Maxwell and Helmholtz systems could be established for some special but important classes of domains [7, 9]. A significant advantage of the quaternionic calculus is that its formulae hold universally for all bounded Lipschitz domains, independently of its particular geometry. Furthermore, one gets very convenient analytic representation formulae as well as rather explicit existence and uniqueness criteria. The application of the quaternionic calculus also leads to explicit expressions for the Lipschitz contraction constant for the fixed-point method solving non-linear problems. Based on the explicit knowledge of the contraction constant one obtains useful a-priori and a posteriori estimates on the iterative approximations.

As shown by Sijue Wu in [24], the quaternionic analysis calculus turned out be a key ingredient in solving fundamental problems related to the well-posedness of the full 3D water wave problem in Sobolev spaces, where the application of well established methods from harmonic and numerical analysis methods did not lead to any success.

About ten years ago, as shown in several papers by P. Cerejeiras, U. Kähler, F. Sommen, and others cf. e.g. [4, 5], these methods were adapted for dealing with the instationary counterparts of these problems in unbounded Lipschitz domains, by means of introducing a parabolic Dirac operator which factorizes the heat operator. To treat the time-dependent case one adds to the canonical basis elements e_1, e_2, e_3 two further basis elements $\mathfrak{f}$ and $\mathfrak{f}^\dagger$ which satisfy $\mathfrak{f}^2 = (\mathfrak{f}^\dagger)^2 = 0$. The additional elements are often called a Witt basis and they allow for the inclusion of the time dimension.

By means of a general positive real $k > 0$ we consider the (modified) parabolic Dirac operator

$$D_{\mathbf{x},t;k}^{\pm} := \sum_{j=1}^{3} e_j \frac{\partial}{\partial x_j} + \mathfrak{f}\frac{\partial}{\partial t} \pm k^2 \mathfrak{f}^\dagger.$$

This operator factorizes the generalized heat operator viz $(D_{\mathbf{x},t;k}^{\pm})^2 = -\Delta \pm k^2 \frac{\partial}{\partial t}$. Similarly to the elliptic case, for this operator one can also introduce adequate analogues of the Teodorescu transform, the regular and singular Cauchy transform and the Bergman projection operator, as proposed in [2, 5, 6, 16]. This adapted operator calculus allows us to treat the time-dependent versions of the PDEs studied earlier over time varying domains in a rather elegant way.

A further recent line of investigation consists in developing possible extensions of this operator calculus to handle such PDE on curved spaces and manifolds. If we want to study for instance weather cast problems, then one appropriate model consists of working with the Navier-Stokes system on the sphere. The latter then involves spherical versions of the Laplacian and the Dirac operator, as proposed for instance by W. Sprößig in [23]. In comparison with the earth radius the atmosphere has a negligible small thickness, so that one deals at first glance with a flow problem on a sphere. To apply the quaternionic operator calculus to the spherical case, one has to adapt the integral operators in a geometric appropriate way, namely the Euclidean Cauchy kernel has to be substituted by its spherical analogue. However, the representation of the solution again consists of the same types of integral operators as introduced in the Euclidean case. One simply has to compute the kernel functions for the new geometry and to replace the Euclidean kernels by the properly adapted versions of these kernels.

In this paper we want to outline how one can adapt the quaternionic operator calculus to study time-dependent Navier-Stokes problems in the more general context of conformally flat spin manifolds that arise by factorizing out some simply connected domain by a discrete Kleinian group. Here, we treat conformally flat spin cylinders an tori as an illustrative example. This underscores the universality of our approach. Furthermore, our approach has the advantage that the representation formulas and criteria can directly be generalized to the n-dimensional setting, just by replacing the quaternionic operators by their corresponding Clifford algebra valued ones, such as suggested in [5] for the Navier-Stokes system, in which all formulas remain with the same structure. We are able to construct the parabolic Cauchy kernel on these manifolds explicitly. This consequently opens the door to apply the iterative computation algorithm to compute the solutions.

2 Preliminaries

2.1 Quaternionic Function Theory

Let $\{e_1, e_2, e_3\}$ denote the standard basis of the Euclidean vector space $\mathbb{R}^3$. To endow the space $\mathbb{R}^3$ with an additional multiplicative structure, we embed it into the Hamiltonian algebra of real quaternions, denoted by $\mathbb{H}$. A quaternion is an element of the form

$$x = x_0 + \mathbf{x} := x_0 + x_1 e_1 + x_2 e_2 + x_3 e_3$$

where $x_0, \ldots, x_3$ are real numbers. x_0 is called the real part of the quaternion x and will be denoted by $\mathrm{Re}(x)$ while $\mathbf{x}$, or $\mathrm{Vec}(x)$, denotes the vector part of x. In the quaternionic setting the standard unit vectors play the role of imaginary units, i.e., we have

$$e_1 e_2 e_3 = e_i^2 = -1, \quad i = 1, 2, 3.$$

The generalized anti-automorphism *conjugation* in $\mathbb{H}$ is defined by

$$\bar{1} = 1, \quad \bar{e_i} = -e_i, \ i = 1, 2, 3, \quad \overline{ab} = \bar{b}\,\bar{a}.$$

The Euclidean norm in $\mathbb{R}^4$ induces a norm on the whole quaternionic algebra as $|a| := \sqrt{\sum_{i=0}^{3} a_i^2}$.

In what follows let $G \subset \mathbb{R}^3$ be a bounded domain with a sufficiently smooth boundary $\Gamma = \partial G$. A quaternionic function $f : G \subset \mathbb{R}^3 \to \mathbb{H}$ has a representation

$$\mathbf{x} \mapsto f(\mathbf{x}) = f_0(\mathbf{x}) + \mathbf{f}(\mathbf{x}) := f_0(\mathbf{x}) + \sum_{i=1}^{3} f_i(\mathbf{x}) e_i,$$

with $\mathbb{R}$−valued components f_i. Properties like continuity, etc. are understood coordinatewisely. Now, the additional multiplicative structure of the quaternions allows to describe all C^1-functions $\mathbf{f} : \mathbb{R}^3 \to \mathbb{R}^3$ that satisfy both div $\mathbf{f} = 0$ and rot $\mathbf{f} = 0$ in a compact form as null-solutions of the three-dimensional Dirac operator

$$\mathbf{D} := \sum_{i=1}^{3} \frac{\partial}{\partial x_i} e_i.$$

This operator is nothing else than the Atiyah-Singer-Dirac operator that arises in a natural way from the Levi-Civita connection in the context of Riemannian spin manifolds. In the Euclidean $3D$-space it simplifies to the form above. More important, the Euclidean Dirac operator coincides with the usual gradient operator when applied to a scalar-valued function. This motivates the study of *monogenic functions*. A real differentiable function $f : G \subset \mathbb{R}^3 \to \mathbb{H}$ is called left quaternionic monogenic, or left quaternionic holomorphic, in G if one has $\mathbf{D}f = 0$ in G. Since the Euclidean Dirac operator factorize up to signal the (3D) Laplace operator, that is $\Delta f = -\mathbf{D}^2 f$, we have that every real component of a left monogenic function is again a harmonic function. Conversely, following e.g. [10], if $f \in C^2$ is a solution of the Laplace operator in G, then in any open ball $B(\tilde{\mathbf{x}}, r) \subset G$ there exist two left monogenic functions, f_0 and f_1, such that $f = f_0 + \mathbf{x} f_1$ holds in $B(\tilde{\mathbf{x}}, r)$. This property allows to treat harmonic functions in terms of null solutions of $\mathbf{D}$. It is also the starting point for the construction of analogues to several well known theorems of complex analysis. For more details on quaternionic functions and operator theory, we refer the reader for instance to [10, 13, 14].

2.2 The Instationary Case

To treat time dependent problems in $\mathbb{R}^3$ we follow the ideas of [5]. First, we introduce two additional basis elements $\mathfrak{f}$ and $\mathfrak{f}^\dagger$ satisfying to

$$\mathfrak{f}\mathfrak{f}^\dagger + \mathfrak{f}^\dagger\mathfrak{f} = 1, \quad \mathfrak{f}^2 = (\mathfrak{f}^\dagger)^2 = 0, \tag{4}$$

and which interact with the existent elements of the basis of $\mathbb{R}^3$ as

$$\mathfrak{f}e_j = e_j\mathfrak{f} = 0, \quad \mathfrak{f}^\dagger e_j = e_j\mathfrak{f}^\dagger = 0. \tag{5}$$

We construct the (dual) parabolic Dirac operators given by

$$D_{\mathbf{x},t}^+ := \mathbf{D} + \mathfrak{f}\frac{\partial}{\partial t} + \mathfrak{f}^\dagger, \quad D_{\mathbf{x},t}^- := \mathbf{D} + \mathfrak{f}\frac{\partial}{\partial t} - \mathfrak{f}^\dagger.$$

We remark that, based on (4) and (5) these operators satisfy $(D_{x,t}^{\pm})^2 = -\Delta \pm \frac{\partial}{\partial t}$, that is to say, they factorize the *heat operator*. Null solutions of the parabolic Dirac operator $D_{x,t}^+$ are called (left) parabolic monogenics (resp. dual parabolic monogenics if solutions of $D_{x,t}^- f = 0$).

Suppose now that G is a space-time varying bounded Lipschitz domain $G \subset \mathbb{R}^3 \times \mathbb{R}^+$. In what follows we define $W_2^{s,l}(G)$ as the parabolic Sobolev space of $L_2(G)$ where s is the regularity parameter with respect to x and l the regularity parameter with respect to t. Using the Stokes theorem we get

$$\int_G \left[\left(g(\mathbf{D} + \mathfrak{f}\partial_t) \right) f + g\left((\mathbf{D} + \mathfrak{f}\partial_t) \right) f \right] dx = \int_\Gamma g d\sigma_{x,t} f,$$

where $d\sigma_{x,t} = \left(\mathbf{D} + \mathfrak{f}\frac{\partial}{\partial t} \right) \rfloor dx dt$ is the contraction of the homogenous operator $\mathbf{D} + \mathfrak{f}\partial_t$ with the volume element $dx dt$. Hence, this leads to the Stokes integral formula involving out parabolic Dirac operators, namely

$$\int_G \left[(g D_{x,t}^-) f + g(D_{x,t}^+ f) \right] dx = \int_\Gamma g d\sigma_{x,t} f. \tag{6}$$

Moreover, the fundamental solution to the dual operator $D_{x,t}^-$ has the form

$$E_-(\mathbf{x}, t) = \frac{H(t) \exp(-\frac{|\mathbf{x}|^2}{4t})}{(2\sqrt{\pi t})^3} \left(\frac{\mathbf{x}}{2t} + \mathfrak{f}(\frac{3}{2t} + \frac{|\mathbf{x}|^2}{4t^2}) - \mathfrak{f}^\dagger \right),$$

where $H(\cdot)$ stands for the usual Heavyside function. Replacing the fundamental solution in (6) we obtain the Borel-Pompeiu integral formula.

Theorem 1 (see [5, 6]) *Let $G \subset \mathbb{R}^3 \times \mathbb{R}^+$ be a bounded Lipschitz domain with a strongly Lipschitz boundary $\Gamma = \partial G$.*
Then for all $u \in W_2^{1,1}(G)$ we have

$$\int_\Gamma E_-(\mathbf{x} - \mathbf{y}, t - \tau) d\sigma_{x,t} u(\mathbf{x}, t) = u(\mathbf{y}, \tau) + \int_G E_-(\mathbf{x} - \mathbf{y}, t - \tau)(D_{x,t}^+ u)(\mathbf{x}, t) d\mathbf{x} dt.$$

Whenever $u \in \text{Ker } D_{x,t}^+$ one obtains the following version of Cauchy's integral formula for parabolic monogenic functions

$$u(\mathbf{y}, \tau) = \int_\Gamma E_-(\mathbf{x} - \mathbf{y}, t - \tau) d\sigma_{x,t} u(\mathbf{x}, t).$$

Again, following the above cited works, we can introduce the parabolic Teodorescu transform and the Cauchy transform by

$$T_G u(\mathbf{y}, \tau) = - \int_G E_-(\mathbf{x} - \mathbf{y}, t - \tau) u(\mathbf{x}, t) d\mathbf{x} dt,$$

$$F_\Gamma u(\mathbf{y}, \tau) = \int_\Gamma E_-(\mathbf{x} - \mathbf{y}, t - \tau) d\sigma_{\mathbf{x},t} u(\mathbf{x}, t).$$

On the one hand we have $D_{\mathbf{x},t}^+ T_G u = u$, that is, the parabolic Teodorescu operator is the right inverse of the parabolic Dirac operator. On the other hand, and analogously to the Euclidean case we can rewrite the Borel-Pompeiu formula in the form

Lemma 1 *Let* $u \in W_2^{1,0}(G)$. *Then*

$$D_{\mathbf{x},t}^+ T_G u = u - F_\Gamma u.$$

The space $L_2(G)$ can be decomposed into the direct sum of the subspace of parabolic monogenics in G and its complement.

Theorem 2 (Hodge decomposition) *Let* $G \subseteq \mathbb{R}^3 \times \mathbb{R}^+$ *be a bounded Lipschitz domain. Then*

$$L_2(G) = \left(L_2(G) \cap \operatorname{Ker} D_{\mathbf{x},t}^+ \right) \oplus D_{\mathbf{x},t}^+ \overset{\circ}{W}_2^{1,1}(G),$$

where $L_2(G) \cap \operatorname{Ker} D_{\mathbf{x},t}^+ =: B(G)$ *is the Bergman space of parabolic monogenic functions, and where* $\overset{\circ}{W}_2^{1,1}(G)$ *is the subspace of all* $f \in W_2^{1,1}(G)$ *with vanishing boundary data.*

Proofs of the above results can be found in [5, 6].

Remark Due to the exponential decrease of the fundamental solution, the operator T_G remains a $L^2(G)$ bounded operator also if $G \subset \mathbb{R}^3 \times \mathbb{R}^+$ is unbounded. The application of the add-on term as proposed in [4] for the Teodorescu transform associated to the usual spatial Dirac operator $\mathbf{D}$ is not necessary in the parabolic setting.

For our purpose we need the more general parabolic Dirac operator, used for instance in [2, 6, 16], having the form

$$D_{\mathbf{x},t,k}^\pm := \mathbf{D} + \mathfrak{f} \frac{\partial}{\partial t} \pm k \mathfrak{f}^\dagger = \sum_{j=1}^3 e_j \frac{\partial}{\partial x_j} + \mathfrak{f} \frac{\partial}{\partial t} \pm k \mathfrak{f}^\dagger$$

for a positive real $k \in \mathbb{R}$. This operator factorizes the second order operator

$$(D_{\mathbf{x},t,k}^\pm)^2 = -\Delta \pm k^2 \frac{\partial}{\partial t}$$

and has very similar properties as the previously ones. Their null-solutions are called parabolic k-monogenic (resp. dual parabolic k-monogenic) functions.

Adapting from [2, 6], the fundamental solution to $D^-_{\mathbf{x},t,k}$ turns out to have the form

$$E(\mathbf{x}, t; k) = \sqrt{k}\,\frac{H(t)\exp(-\frac{k|\mathbf{x}|^2}{4t})}{(2\sqrt{\pi t})^3}\left(\frac{k}{2t}\mathbf{x} + \mathfrak{f}(\frac{3}{2t} + \frac{k|\mathbf{x}|^2}{4t^2}) + k^2\mathfrak{f}^\dagger\right).$$

In what follows $\mathbf{P} : L_2(G) \to B(G) := L_2(G) \cap \operatorname{Ker} D^+_{\mathbf{x},t;k}$ denotes the orthogonal Bergman projection while $\mathbf{Q} : L_2(G) \to D^+_{\mathbf{x},t}\overset{\circ}{W}_2^{1,1}(G)$ stands for the projection
into the complementary space in all that follows. One has $\mathbf{Q} = \mathbf{I} - \mathbf{P}$. Here $\mathbf{I}$ stands
for the identity operator.

The Bergman space of parabolic k-monogenic functions is a Hilbert space
with a uniquely defined reproducing kernel function, the so-called the parabolic
k-monogenic Bergman kernel denoted by $B(\mathbf{x}, \mathbf{y}; t, \tau)$. The orthogonal Bergman
projection $\mathbf{P} : L_2(G) \to B(G)$ is given by the convolution with the Bergman kernel

$$(\mathbf{P}u)(\mathbf{x}, t) = \int_G B(\mathbf{x}, \mathbf{y}; t, \tau)u(\mathbf{y}, \tau)d\mathbf{y}d\tau, \quad u \in L_2(G).$$

In particular, one has $(\mathbf{P}u)(\mathbf{x}, t) = u(\mathbf{x}, t)$ for all $u \in B(G)$.

3 The Navier-Stokes Equations Shortly Revisited in Quaternions

In the classical vector analysis calculus the in-stationary Navier-Stokes equations
have the form (again, we assume here viscosity $\nu = 1$)

$$-\Delta \mathbf{u} + \frac{\partial \mathbf{u}}{\partial t} + (\mathbf{u}\,\mathrm{grad})\,\mathbf{u} + \mathrm{grad}\,p = \mathbf{f}, \quad \text{in } G \tag{7}$$

$$\mathrm{div}\,\mathbf{u} = 0, \quad \text{in } G \tag{8}$$

$$\mathbf{u} = \mathbf{0}, \quad \text{on } \partial G \tag{9}$$

To apply the quaternionic integral operator calculus to solve these equations one first
expresses this system in the quaternionic language, as done in [5].

First we recall that for a time independent quaternionic function of type

$$\mathbf{f} : \mathbb{R}^3 \to \mathbb{R}^3, \quad \text{with } \mathbf{x} \mapsto \mathbf{f}(\mathbf{x})$$

we have

$$\mathbf{D}\mathbf{f} = \mathrm{rot}\,\mathbf{f} - \mathrm{div}\,\mathbf{f}.$$

Hence, the divergence of a vector field $\mathbf{f}$ can be expressed as $\mathrm{div}\,\mathbf{f} = \mathrm{Re}(\mathbf{D}\mathbf{f})$.

In a similar way, for a scalar valued function $p : \mathbb{R}^3 \to \mathbb{R},\ \mathbf{x} \mapsto p(\mathbf{x})$, we have

$$\mathbf{D}p = \text{grad } p.$$

Finally, we recall that the three dimensional Euclidean Laplacian $\Delta = \sum_{i=1}^{3} \frac{\partial^2}{\partial x_i^2}$ can be expressed in terms of the Dirac operator as $\Delta = -\mathbf{D}^2$.

Next, we assume that the vector-field is time-dependent, that is, $\mathbf{u} = \mathbf{u}(\mathbf{x}, t)$. Applying the formulas from the preceding section we can express the heat operator $-\Delta\mathbf{u} + \frac{\partial\mathbf{u}}{\partial t}$ in the form

$$-\Delta\mathbf{u} + \frac{\partial\mathbf{u}}{\partial t} = (D_{\mathbf{x},t}^+)^2\mathbf{u}.$$

Thus, the original system for a time-dependent vector-field $\mathbf{u} = \mathbf{u}(\mathbf{x}, t)$ can be reformulated in the following way:

$$(D_{\mathbf{x},t}^+)^2\mathbf{u} + \text{Re}(\mathbf{u}\,\mathbf{D})\,\mathbf{u} + \mathbf{D}\,p = \mathbf{f} \text{ in } G \qquad (10)$$

$$\text{Re}(\mathbf{Du}) = 0 \text{ in } G \qquad (11)$$

$$\mathbf{u} = \mathbf{0}, \text{ at } \partial G. \qquad (12)$$

The strategy for the resolution of this system is to apply the previously introduced hypercomplex integral operators in order to get iterative formulas for the velocity $\mathbf{u}$ and the pressure p.

4 The Linear Case

In this section we briefly recall how the quaternionic calculus can be applied to set up analytic solutions for the special case in which the convective term $(\mathbf{u}\,\text{grad})\,\mathbf{u}$ is negligibly small. Hence, we assume p and the external source $\mathbf{f}$ in $L_2(G)$ and $\mathbf{u} \in \overset{\circ}{W}_2^{1,1}(G)$.

Under these assumptions the instationary viscous Navier-Stokes equations take the simplified form

$$(D_{\mathbf{x},t}^+)^2\mathbf{u} + \mathbf{D}\,p = \mathbf{f} \text{ in } G \qquad (13)$$

$$\text{Re}(\mathbf{Du}) = 0 \text{ in } G \qquad (14)$$

$$\mathbf{u} = \mathbf{0} \text{ at } \partial G. \qquad (15)$$

The velocity $\mathbf{u}$ and the pressure p can now be computed from this system using the (modified) parabolic Teodorescu operator. Applying the parabolic Teodorescu operator to (13) leads to the equation

$$(T_G D_{\mathbf{x},t}^+)(D_{\mathbf{x},t}^+\mathbf{u}) + T_G\mathbf{D}p = T_G(\mathbf{f}). \qquad (16)$$

Now, we apply Lemma 1 (Borel-Pompeiu formula) to (16). This leads to

$$(D_{x,t}^+ \mathbf{u} - F_\Gamma D_{x,t}^+ \mathbf{u}) + T_G \mathbf{D} p = T_G(\mathbf{f}). \tag{17}$$

Using the orthogonal Bergman projector $\mathbf{Q}$ yields

$$(\mathbf{Q} D_{x,t}^+ \mathbf{u} - \mathbf{Q} F_\Gamma D_{x,t}^+ \mathbf{u}) + \mathbf{Q} T_G \mathbf{D} p = \mathbf{Q} T_G(\mathbf{f}). \tag{18}$$

Since the Cauchy integral operator maps $L_2(G)$ onto $L_2(G) \cap \mathrm{Ker}\, D_{x,t}^+$ and $\mathbf{u} \in \overset{\circ}{W}_2^{1,1}(G)$ we get that $F_\Gamma D_{x,t}^+ \mathbf{u}$ is a left parabolic monogenic function that is, $\mathbf{Q} F_\Gamma D_{x,t}^+ \mathbf{u} = 0$.

Therefore, Eq. (18) simplifies to

$$\mathbf{Q} D_{x,t}^+ \mathbf{u} + \mathbf{Q} T_G \mathbf{D} p = \mathbf{Q} T_G(\mathbf{f}). \tag{19}$$

At this point we remark that T_G is the right inverse to $D_{x,t}^+$ but not to $\mathbf{D}$!

We apply again the Teodorescu transform to Eq. (19):

$$T_G \mathbf{Q} D_{x,t}^+ \mathbf{u} + T_G \mathbf{Q} T_G \mathbf{D} p = T_G \mathbf{Q} T_G(\mathbf{f}). \tag{20}$$

First, we observe that $T_G \mathbf{Q} D_{x,t}^+ \mathbf{u} = T_G D_{x,t}^+ \mathbf{u}$, because $D_{x,t}^+ \mathbf{u} \in im(\mathbf{Q})$. Applying again Lemma 1 leads to

$$\mathbf{u} - F_\Gamma \mathbf{u} + T_G \mathbf{Q} T_G \mathbf{D} p = T_G \mathbf{Q} T_G(\mathbf{f}). \tag{21}$$

Since $\mathbf{u}|_\Gamma = \mathbf{0}$, we get that $F_\Gamma \mathbf{u}$ vanishes and we do obtain the following representation formula for the velocity field $\mathbf{u}$:

$$\mathbf{u} = T_G \mathbf{Q} T_G (\mathbf{f} - \mathbf{D} p). \tag{22}$$

The pressure p can be obtained from the continuity Eq. (14). Indeed, inserting the solution $\mathbf{u}$ obtained in (22) into (14) leads to

$$Re(\mathbf{Q} T_G \mathbf{D} p) = Re(\mathbf{Q} T_G \mathbf{f}). \tag{23}$$

Thus, we have obtained the following representation formulas for the solutions of the instationary viscous Navier-Stokes equations in the case of a negligibly small convective term:

Theorem 3 (Representation theorem) *Suppose that $p \in L_2(G)$, $\mathbf{u} \in \overset{\circ}{W}_2^{1,1}(G)$. Then the solutions can be represented in the form*

$$Re(\mathbf{Q} T_G \mathbf{D} p) = Re(\mathbf{Q} T_G \mathbf{f}) \tag{24}$$

$$\mathbf{u} = T_G \mathbf{Q} T_G \mathbf{f} - T_G \mathbf{Q} T_G \mathbf{D} p. \tag{25}$$

The pressure is uniquely determined from (24) up to a constant. Given the solution for the pressure p, (25) gives the solution for $\mathbf{u}$. Hence, the original system is solvable by application of the integral operators T_G and $\mathbf{Q}$.

Both the Teodorescu and Cauchy integral operators have a universal integral kernel for all bounded domains, namely the Cauchy kernel $E = E(\mathbf{x} - \mathbf{y}; t - \tau)$. Also, the Bergman projectors can be expressed by the algebraic relation

$$\mathbf{P} = F_\Gamma (tr_\Gamma T_G F_\Gamma)^{-1} tr_\Gamma T_G,$$

where tr_Γ is the usual trace operator, or the restriction to the boundary of the domain. Furthermore, the above scheme is extendable to the case of k–monogenics, that is, when (13) is given as $(\mathbf{D}_{\mathbf{x},t;k})^2 \mathbf{u} + \mathbf{D}p = \mathbf{f}$. See [5, 13] for more details.

5 The Case of a Non Negligenciable Convective Term

Now, we turn our attention to the more complicated case in which the flow is still viscous ($\nu = 1$) but the non-linear convective term $(\mathbf{u}\ \mathrm{grad})\mathbf{u} = \mathrm{Re}(\mathbf{u}\mathbf{D})\mathbf{u}$ is no longer negligibly small.

First, we observe that the reasoning and arguments used in Sect. 4 are still valid. Hence, we get the following equations for the velocity $\mathbf{u}$ and pressure p, that is,

$$\mathbf{u} = T_G \mathbf{Q} T_G \Big[\mathbf{f} - \mathrm{Re}(\mathbf{u}\mathbf{D})\mathbf{u} \Big] - T_G \mathbf{Q} T_G \mathbf{D}p, \tag{26}$$

and, from inserting this solution into the continuity Eq. (14), we obtain

$$\mathrm{Re}(\mathbf{Q}T_G \mathbf{D}p) = \mathrm{Re}\Big[\mathbf{Q}T_G\ (\mathbf{f} - \mathrm{Re}(\mathbf{u}\mathbf{D})\mathbf{u}) \Big]. \tag{27}$$

Now, we apply the following fixed point algorithm in order to iteratively compute both solution $\mathbf{u}$ and pressure p, departing from an arbitrary $\mathbf{u}_0$ (for the time being, no conditions will be imposed here):

$$\mathrm{Re}(\mathbf{Q}T_G \mathbf{D}p_n) = \mathrm{Re}\Big[\mathbf{Q}T_G \big(\mathbf{f} - \mathrm{Re}(\mathbf{u}_{n-1}\mathbf{D})\mathbf{u}_{n-1} \big) \Big],$$

$$\mathbf{u}_n = T_G \mathbf{Q} T_G \Big[\mathbf{f} - \mathrm{Re}(\mathbf{u}_{n-1}\mathbf{D})\mathbf{u}_{n-1} \Big] - T_G \mathbf{Q} T_G \mathbf{D}p_n$$

for $n = 1, 2, \ldots$

The following lemma (c.f. [5]) gives the conditions under which the above proposed fixed point algorithm does converge to a unique solution $(\mathbf{u}, p)$:

Lemma 2 *Suppose that* $\mathbf{u} \in \overset{\circ}{W}_2^{s,l}(G) \cap Ker\mathbf{D}$, *with* $s, l \geq 1$, *and* $p \in L_2(G)$ *are pairwise solutions of (26) and (27). Then, the following estimate holds:*

$$\|D_{x,t}^+ \mathbf{u}\|_2 + \|\mathbf{Q}p\|_2 \le \sqrt{2}\|T_G M(\mathbf{u})\|_2, \tag{28}$$

where $M(\mathbf{u}) := Re(\mathbf{u}D)\mathbf{u} - \mathbf{f}$ *and* $\|\cdot\|_2$ *stands for the* L_2-*norm.*

In fact, for $p \in W_2^{1,1}(G)$ we have

$$T_G \mathbf{Q} T_G \mathbf{D}p = T_G \mathbf{Q}(p - F_\Gamma p) = T_G \mathbf{Q}p,$$

since $F_\Gamma p$ is in im$\mathbf{P}$. Applying $D_{x,t}^+$ to this equation gives (recall that $D_{x,t}^+$ is a left inverse for T_G)

$$D_{x,t}^+(T_G \mathbf{Q} T_G \mathbf{D}p) = \mathbf{Q}p.$$

Since $W_2^{1,1}(G)$ is dense in $L_2(G)$ this leads to $D_{x,t}^+(T_G \mathbf{Q} T_G \mathbf{D}p) = \mathbf{Q}p$ for all $p \in L_2(G)$, and we obtain from (27)

$$D_{x,t}^+ \mathbf{u} = \mathbf{Q} T_G M(\mathbf{u}) - \mathbf{Q}p.$$

By the orthogonality between $D_{x,t}^+\mathbf{u}$ and $\mathbf{Q}p$, we have

$$\|D_{x,t}^+ \mathbf{u}\|_2 + \|\mathbf{Q}p\|_2 \le \sqrt{2}\|\mathbf{Q}T_G M(\mathbf{u})\|_2 = \sqrt{2}\|T_G M(\mathbf{u})\|_2.$$

Starting with $\mathbf{u}_0 \in \overset{\circ}{W}_2^{1,1}(G)$ we generate the iteration pairs $(p_n, \mathbf{u}_n) \in L_2(G) \times \overset{\circ}{W}_2^{1,1}(G)$. Moreover, by (27) we get

$$\|\mathbf{u}_n - \mathbf{u}_{n-1}\|_{W_2^{1,1}} \le \|T_G \mathbf{Q} T_G[M(\mathbf{u}_{n-1}) - M(\mathbf{u}_{n-2})]\|_{W_2^{1,1}} + \|T_G \mathbf{Q}(p_n - p_{n-1})\|_{W_2^{1,1}}$$
$$\le 2C_1 \|M(\mathbf{u}_{n-1}) - M(\mathbf{u}_{n-2})\|_{W_2^{-1,-1}},$$

where $C_1 := \|T_G \mathbf{Q} T_G\|$.

Moreover, according to [4] Lemma 4.1 for all $\mathbf{u} \in \overset{\circ}{W}_2^{1,1}(G)$ there exists a constant C_2 such that

$$\|Re(\mathbf{u}D)\mathbf{u}\|_{W_2^{-1,-1}} \le C_2 \|\mathbf{u}\|_{W_2^{1,1}}^2,$$

so that the previous estimate becomes

$$\|\mathbf{u}_n - \mathbf{u}_{n-1}\|_{W_2^{1,1}} \le 2C_1 C_2 \left(\|\mathbf{u}_{n-1}\|_{W_2^{1,1}} + \|\mathbf{u}_{n-2}\|_{W_2^{1,1}} \right) \|\mathbf{u}_{n-1} - \mathbf{u}_{n-2}\|_{W_2^{1,1}}.$$

Next, we need to prove that the energy of our solutions $\mathbf{u}_n$ decreases, that is to say, $\|\mathbf{u}_n\|_{W_2^{1,1}} \le \|\mathbf{u}_{n-1}\|_{W_2^{1,1}}$. First, we observe that

$$\|\mathbf{u}_n\|_{W_2^{1,1}} \le \|T_G \mathbf{Q} T_G \mathbf{u}_{n-1}\|_{W_2^{1,1}} + \|T_G \mathbf{Q} p_n\|_{W_2^{1,1}}$$
$$\le 2C_1 C_2 \|\mathbf{u}_{n-1}\|_{W_2^{1,1}}^2 + 2C_1 \|\mathbf{f}\|_2.$$

Hence, $\|\mathbf{u}_n\|_{W_2^{1,1}} \leq \|\mathbf{u}_{n-1}\|_{W_2^{1,1}}$ whenever

$$2C_1C_2\|\mathbf{u}_{n-1}\|_{W_2^{1,1}}^2 + 2C_1\|\mathbf{f}\|_2 \leq \|\mathbf{u}_{n-1}\|_{W_2^{1,1}}$$

which leads to

$$\left(\|\mathbf{u}_{n-1}\|_{W_2^{1,1}} - \frac{1}{4C_1C_2}\right)^2 \leq \frac{1}{16C_1^2C_2^2} - \frac{1}{C_2}\|\mathbf{f}\|_2.$$

Now, if $\|\mathbf{f}\|_2 \leq \frac{1}{16C_1^2C_2}$, then the previous in-equation can be written as

$$\frac{1}{4C_1C_2} - W \leq \|\mathbf{u}_{n-1}\|_{W_2^{1,1}} \leq \frac{1}{4C_1C_2} + W,$$

where $W := \sqrt{\frac{1}{16C_1^2C_2^2} - \frac{1}{C_2}\|\mathbf{f}\|_2}$. This finally leads to an estimate on the Lipschitz constant

$$\|\mathbf{u}_n - \mathbf{u}_{n-1}\|_{W_2^{1,1}} \leq 2C_1C_2\left(\|\mathbf{u}_{n-1}\|_{W_2^{1,1}} + \|\mathbf{u}_{n-2}\|_{W_2^{1,1}}\right)\|\mathbf{u}_{n-1} - \mathbf{u}_{n-2}\|_{W_2^{1,1}}$$

$$\leq \left(1 + 2C_1C_2W\right)\|\mathbf{u}_{n-1} - \mathbf{u}_{n-2}\|_{W_2^{1,1}},$$

of the form

$$L := 1 - 4C_1C_2W < 1.$$

Summarizing

Theorem 4 (cf. [5] p. 1723).
The iteration method converges for each starting point $\mathbf{u}_0 \in \overset{\circ}{W}_2^{1,1}(G) \cap ker\mathbf{D}$ *with*

$$\|\mathbf{u}_0\|_{W_2^{1,1}} \leq \min\left(\frac{1}{2C_1C_2}, \frac{1}{4C_1C_2} + W\right)$$

and $W := \sqrt{\frac{1}{16C_1^2C_2^2} - \frac{1}{C_2}\|\mathbf{f}\|_2}$.

Remark we get the pressure p up to an additive constant when G is bounded, and we get uniqueness of p when G is unbounded. Also, and as explained in [1] in the time independent case we can replace the Teodorescu transform by a simpler primitivation operator whose evaluation requires less computational steps.

6 The Navier-Stokes Equations in the More General Context of Some Conformally Flat Spin 3-Manifolds

One further advantage of using the quaternionic operator calculus consists in the fact that the previous results can easily be carried over to the treatment of analogous boundary value problems on conformally flat spin manifolds. As a simple example, consider the Euclidean space $\mathbb{R}^3$. This is due to the fact that the formulas presented in the previous sections have geometrically a very universal character.

Recalling for example from the classical paper [19] a conformally flat 3-manifold is a Riemannian 3-manifold that has a vanishing Weyl tensor. In dimensions $n \geq 3$ these are exactly those Riemannian manifolds that have atlasses whose transition functions are Möbius transformations.

As also pointed out in [19], one way of constructing examples of conformally flat manifolds is to factor out a subdomain U of either the sphere S^3 or $\mathbb{R}^3$ by a Kleinian subgroup Γ of the Möbius group, where Γ acts totally discontinuously on U. This gives rise to the conformally flat manifold U/Γ. In the original paper by N.H. Kuiper it is shown that the universal cover of a conformally flat manifold admits a development (i.e. a local conformal diffeomorphism) into S^3. The class of conformally flat manifolds of the form U/Γ are exactly those for which this development is a covering map $\tilde{U} \to U \subset S^3$.

Examples of such manifolds are for example 3-tori, cylinders, real projective space and the hyperbolic manifolds $H^+/\Gamma_p[N]$ for an integer $N \geq 2$, where $H^+ := \{\mathbf{x} \in \mathbb{R}^3 \mid x_3 > 0\}$ and where $\Gamma_p[N]$ is a principal congruence arithmetic subgroup of level N of the hypercomplex modular group Γ_p. The latter one is generated by the Kelvin inversion (i.e. the reflection at the unit sphere $\mathbf{x} \mapsto -\mathbf{x}/|\mathbf{x}|^2$) and by the translation operations $\mathbf{x} \mapsto \mathbf{x} + e_i$ for $i \leq 2$. This group generalizes the group $PSL(2, \mathbb{Z})$ to higher dimensions. The quotient space of H^+ with a principal subgroup $\Gamma_p[N]$ is indeed a manifold for $N \geq 2$, because $\Gamma_p[N]$ is torsion-free whenever $N \geq 2$. $\Gamma_p[N]$ consists of those matrices $\begin{pmatrix} a & b \\ c & d \end{pmatrix}$ from Γ_p where the entries satisfy the arithmetic conditions $a - 1, b, c, d - 1 \equiv 0 \mod \mathbb{Z} + \mathbb{Z}e_1 + \mathbb{Z}e_2 + \mathbb{Z}e_3$. For more profound details on these groups and properties we refer the reader to [3].

In order to generalize the previous results to the context of analogous instationary boundary value problems on conformally manifolds, we only need to introduce the properly adapted analogues of the parabolic Dirac operator, and of the other hypercomplex integral operators on these manifolds. So, the main goal consists of constructing the kernel functions explicitly. From the geometric point of view, one is particularly interested in those conformally flat manifolds that have a spin structure, that means that one can construct at least one spinor bundle over such a manifold. These are called conformally spin manifolds. Often one can construct more than one spinor bundle over a spin manifold which leads to the consideration of spinor sections, in our case quaternionic spinor sections. For the geometric background we refer to [20].

In this paper, we restrict ourselves to explain the method at the simplest nontrivial example dealing with conformally flat spin 3-tori with inequivalent spinor bundles. After this, it becomes clear how to carry over our results to other examples of conformally flat (spin) manifolds that are constructed by factoring out a simply connected domain by a discrete Kleinian group, such as those mentioned above.

To start, let $\Omega := \mathbb{Z}e_1 + \mathbb{Z}e_2 + \mathbb{Z}e_3$ be the standard lattice in $\mathbb{R}^3$. Then the topological quotient space $\mathbb{R}^3/\Omega$ is a 3-dimensional conformally flat torus denoted by T_3, over which one can construct a number of conformally inequivalent spinor bundles over T_3.

We recall that in general different spin structures on a spin manifold M are detected by the number of distinct homomorphisms from the fundamental group $\Pi_1(M)$ to the group $\mathbb{Z}_2 = \{0, 1\}$. In this case we have that $\Pi_1(T_3) = \mathbb{Z}^3$. There are two homomorphisms of $\mathbb{Z}$ to $\mathbb{Z}_2$. The first one is $\theta_1 : \mathbb{Z} \to \mathbb{Z}_2 : \theta_1(n) = 0 \mod 2$ while the second one is the homomorphism $\theta_2 : \mathbb{Z} \to \mathbb{Z}_2 : \theta_2(n) = 1 \mod 2$. Consequently there are 2^3 distinct spin structures on T_3. T_3 is a simple example of a Bieberbach manifold.

We shall now give an explicit construction for some of these spinor bundles over T_3. All the others are constructed similarly. First let l be an integer in the set $\{1, 2, 3\}$, and consider the sublattice $\mathbb{Z}^l = \mathbb{Z}e_1 + \cdots + \mathbb{Z}e_l$ where $(0 \leq l \leq 3)$. In the case $l = 0$ we simply have $\mathbb{Z}^0 := \emptyset$. There is also the remainder lattice $\mathbb{Z}^{3-l} = \mathbb{Z}e_{l+1} + \cdots + \mathbb{Z}e_3$. In this case $\mathbb{Z}^3 - \{\underline{m} + \underline{n} : \underline{m} \in \mathbb{Z}^l \text{ and } \underline{n} \in \mathbb{Z}^{3-l}\}$. Suppose now that $\underline{m} = m_1 e_1 + \cdots + m_l e_l$. Let us now make the identification $(\mathbf{x}, X)$ with $(\mathbf{x} + \underline{m} + \underline{n}, (-1)^{m_1 + \cdots + m_l} X)$ where $\mathbf{x} \in \mathbb{R}^3$ and $X \in \mathbb{H}$. This identification gives rise to a quaternionic spinor bundle $E^{(l)}$ over T_3.

Notice that $\mathbb{R}^3$ is the universal covering space of T_3. Consequently, there exists a well-defined projection map $p : \mathbb{R}^3 \to T_3$. As explained for example in [15] every 3-fold periodic resp. anti-periodic open set $U \subset \mathbb{R}^3$ and every 3-fold periodic resp. anti-periodic section $f : U' \to E^{(l)}$, satisfying $f(\mathbf{x}) = (-1)^{m_1 + \cdots + m_l}(\mathbf{x} + \omega)$ for all $\omega \in \mathbb{Z}^l \oplus \mathbb{Z}^{3-l}$, descends to a well-defined open set $U' = p(U) \subset T_3$ (associated with the chosen spinor bundle) and a well-defined spinor section $f' := p(f) : U' \subset T_3 \to E^{(l)} \subset \mathbb{H}$, respectively. The projection map $p : \mathbb{R}^3 \to T_3$ induces well-defined toroidal modified parabolic Dirac operators on $T_3 \times \mathbb{R}^+$ by $p(D_{\mathbf{x},t,k}^\pm) =: \mathscr{D}_{\mathbf{x},t,k}^\pm$ acting on spinor sections of $T_3 \times \mathbb{R}^+$. Sections defined on open sets U of $T_3 \times \mathbb{R}^+$ are called toroidal k-left parabolic monogenic if $\mathscr{D}_{\mathbf{x},t,k}^\pm s = 0$ holds in U. By $\tilde{D} := p(\mathbf{D})$ we denote the projection of the time independent Euclidean Dirac operator to the torus T_3.

The projections of the 3-fold (anti-)periodization of the function $E(\mathbf{x}, t; k)$ denoted by

$$\mathscr{E}(\mathbf{x}, t; k) := \sum_{\omega \in \mathbb{Z}^3 \oplus \mathbb{Z}^{3-l}} (-1)^{m_1 + \cdots + m_l} E(\mathbf{x} + \omega, t; k)$$

provides us with the fundamental section to the toroidal parabolic modified Dirac operator $\mathscr{D}_{\mathbf{x},t,k}^\pm$ acting on the corresponding spinor bundle of the torus T_3. From the function theoretical point of view the function $\mathscr{E}(\mathbf{x}, t; k)$ can be regarded as the canonical generalization of the classical elliptic Weierstraß $\wp$-function to the context of the modified Dirac operator $D_{\mathbf{x},t,k}^+$ in three dimensions.

To show that this expression is well-defined we have to prove the convergence of the series. So, the main task is show

Theorem 5 *The series*

$$\mathscr{E}(\mathbf{x}, t; k) = \sum_{\omega \in \mathbb{Z}^3 \oplus \mathbb{Z}^{3-l}} (-1)^{m_1 + \cdots + m_l} E(\mathbf{x} + \omega, t; k)$$

converges uniformly on any compact subset of $\mathbb{R}^3 \times \mathbb{R}^+$.

Proof We decompose the total lattice $\mathbb{Z}^3$ into the the the following union of lattice points $\Omega = \bigcup_{m=0}^{+\infty} \Omega_m$ where

$$\Omega_m := \{\omega \in \mathbb{Z}^3 \mid |\omega|_{max} = m\}.$$

We further consider the following subsets of this lattice

$$L_m := \{\omega \in \mathbb{Z}^3 \mid |\omega|_{max} \leq m\}.$$

By a direct counting argument one observes the set L_m contains exactly $(2m+1)^3$ many points. Hence, the cardinality of Ω_m is $\sharp\Omega_m = (2m+1)^3 - (2m-1)^3$. The Euclidean distance between the set Ω_{m+1} and Ω_m has the value $d_m := dist_2(\Omega_{m+1}, \Omega_m) = 1$.

To show the normal convergence of the series, let us consider an arbitrary compact subset $\mathscr{K} \subset \mathbb{R}^3$. Let $t > 0$ be an arbitrary but fixed value. Then there exists a positive real $r \in \mathbb{R}$ such that all $\mathbf{x} \in \mathscr{K}$ satisfy $|\mathbf{x}|_{max} \leq |\mathbf{x}|_2 < r$. Suppose now that $\mathbf{x}$ is a point of $\mathscr{K}$. To show the normal convergence of the series we may leave out without loss of generality a finite set of lattice points. So, we retrict ourselves to extend only the summation over those lattice points that satisfy $|\omega|_{max} \geq [r] + 1$. In view of

$$|\mathbf{x} + \omega|_2 \geq |\omega|_2 - |\mathbf{x}|_2 \geq |\omega|_{max} - |\mathbf{x}|_2 = m - |\mathbf{x}|_2 \geq m - r,$$

for $\omega \in \Omega_m$, we obtain

$$\sum_{m=[r]+1}^{+\infty} \sum_{\omega \in \Omega_m} |E(\mathbf{x}, t; k)(\mathbf{x} + \omega)|_2$$

$$\leq \frac{k}{(2\sqrt{\pi t})^3} \sum_{m=[r]+1}^{+\infty} \sum_{\omega \in \Omega_m} \exp(-k|\mathbf{x} + \omega|_2/4t)\Big(\frac{k}{2t}|\mathbf{x} + \omega|_2 + \mathfrak{f}(\frac{3}{2t} + \frac{k|\mathbf{x} + \omega|_2^2}{4t^2}) + k\mathfrak{f}^\dagger\Big)$$

$$\leq \frac{k}{(2\sqrt{\pi t})^3} \sum_{m=[r]+1}^{+\infty} \Big([(2m+1)^3 - (2m-1)^3](\frac{k(m-r)}{2t} + \mathfrak{f}(\frac{3}{2t} + \frac{k(m-r)^2}{4t^2}) + k\mathfrak{f}^\dagger)$$

$$\times \exp(\frac{-k(m-r)^2}{4t})\Big),$$

because $m - r \geq [r] + 1 - r > 0$. This sum clearly is absolutely uniformly convergent because of the decreasing exponent (remember $k > 0$) which dominates the polynomial expressions in m. Hence, the series

$$\mathcal{E}(\mathbf{x}, t; k) := \sum_{\omega \in \mathbb{Z}^l \oplus \mathbb{Z}^{3-l}} (-1)^{m_1 + \cdots + m_l} E(\mathbf{x} + \omega, t; k),$$

which can be rewritten as

$$\mathcal{E}(\mathbf{x}, t; k) := \sum_{m=0}^{+\infty} \sum_{\omega \in \Omega_m} (-1)^{m_1 + \cdots + m_l} E(\mathbf{x} + \omega, t; k),$$

converges normally on $\mathbb{R}^3 \times \mathbb{R}^+$. Since $E(\mathbf{x} + \omega, t; k)$ belongs to Ker $D_{\mathbf{x}, t, k}^+$ in each $(\mathbf{x}, t) \in \mathbb{R}^3 \times \mathbb{R}^+$ the series $\mathcal{E}(\mathbf{x}, t; k)$ satisfies $D_{\mathbf{x}, t, k}^+ \mathcal{E}(\mathbf{x}, t; k) = 0$ in each $\mathbf{x} \in \mathbb{R}^3 \times \mathbb{R}^+$. ∎

Obviously, by a direct rearrangement argument, one obtains that

$$\mathcal{E}(\mathbf{x}, t; k) = (-1)^{m_1 + \cdots + m_l} \mathcal{E}(\mathbf{x} + \omega, t; k) \quad \forall \omega \in \Omega$$

which shows that the projection of this kernel correctly descends to a section with values in the spinor bundle $E^{(l)}$. The projection $p(\mathcal{E}(\mathbf{x}, t; k))$ denoted by $\tilde{\mathcal{E}}(\mathbf{x}, t; k)$ is the fundamental section of the toroidal modified parabolic Dirac operator $\tilde{D}_{\mathbf{x}, t, k}^+$. For a time-varying Lipschitz domain $G \subset T_3 \times \mathbb{R}^+$ with a strongly Lipschitz boundary Γ, we can now similarly introduce the Teodorescu and Cauchy transform for toroidal k-monogenic parabolic quaternionic spinor valued sections by

$$\tilde{T}_G u(\mathbf{y}, t_0) = - \int_G \tilde{\mathcal{E}}(\mathbf{x} - \mathbf{y}, t - t_0; k) u(\mathbf{x}, t) dV dt$$

$$\tilde{F}_\Gamma u(\mathbf{y}, t_0) = \int_\Gamma \tilde{\mathcal{E}}(\mathbf{x} - \mathbf{y}, t - t_0; k) d\sigma_{\mathbf{x}, t} u(\mathbf{x}, t).$$

Next, the associated Bergman projection can be introduced by

$$\tilde{\mathbf{P}} = \tilde{F}_\Gamma (tr_\Gamma \tilde{T}_G \tilde{F}_\Gamma)^{-1} tr_\Gamma \tilde{T}_G.$$

and $\tilde{\mathbf{Q}} := \tilde{\mathbf{I}} - \tilde{\mathbf{P}}$.

Adapting from [8, 15] we obtain a direct analogy of Theorem 1, Lemma 1 and Theorem 2 on these conformally flat 3-tori using these toroidal versions $\tilde{T}_G$, $\tilde{F}_\Gamma$ and $\tilde{\mathbf{P}}$ of operators introduced in Sect. 2. Suppose next that we have to solve a Navier-Stokes problem of the form (1)–(3) within a Lipschitz domain $G \subset T_3 \times \mathbb{R}^+$ with values in the spinor bundle $E^{(l)} \times \mathbb{R}^+$. Then we can compute its solutions by simply applying the following adapted iterative algorithm

$$\mathbf{u}_n = \tilde{T}_G \tilde{Q} \tilde{T}_G \Big[\mathbf{f} - \mathrm{Re}(\mathbf{u}_{n-1} \tilde{D}) \mathbf{u}_{n-1} \Big] - \tilde{T}_G \tilde{Q} \tilde{T}_G \tilde{D} p_n$$

$$\mathrm{Re}(\tilde{Q} \tilde{T}_G \tilde{D} p_n) = \mathrm{Re}\Big[\tilde{Q} \tilde{T}_G \mathbf{f} - \mathrm{Re}(\mathbf{u}_{n-1} \tilde{D}) \mathbf{u}_{n-1} \Big]$$

In the same flavor one obtains a direct analogy of Theorems 3 and 4 in this context.

Now it becomes clear how this approach even carries over to more general conformally flat spin manifolds that arise by factoring out a simply connected domain U by a discrete Kleinian group Γ. The Cauchy-kernel is constructed by the projection of the Γ-periodization (involving eventually automorphy factors like in [3]) of the fundamental solution $E(\mathbf{x}; t; k)$. With this fundamental solution we construct the corresponding integral operators on the manifold.

In terms of these integral operators, we can express the solutions of the corresponding Navier-Stokes boundary value problem on these manifolds, simply by replacing the usual hypercomplex integral operators by its adequate "periodic" analogies on the manifold. This again underlines the very universal character of our approach to treat the Navier-Stokes equations but also many other complicated elliptic, parabolic, hypoelliptic and hyperbolic PDE systems with the quaternionic operator calculus using Dirac operators.

Furthermore, the representation formulas and results also carry directly over to the n-dimensional case in which one simply replaces the corresponding quaternionic operators by Clifford algebra valued operators, such as suggested in [5, 8].

Acknowledgements The work of the third author is supported by the project *Neue funktionentheoretische Methoden für instationäre PDE*, funded by Programme for Cooperation in Science between Portugal and Germany, DAAD-PPP Deutschland-Portugal, Ref: 57340281. The work of the first and second authors is supported via the project "New Function Theoretical Methods in Computational Electrodynamics" approved under the agreement Acções Integradas Luso-Alemãs DAAD-CRUP, ref. A-15/17, and by Portuguese funds through the CIDMA - Center for Research and Development in Mathematics and Applications, and the Portuguese Foundation for Science and Technology ("FCT–Fundação para a Ciência e a Tecnologia"), within project UID/MAT/0416/2013.

References

1. Bahmann, H., Gürlebeck, K., Shapiro, M., Sprößig, W.: On a modified teodorescu transform. Integral Transf. Spec. Funct. **12**(3), 213–226 (2001)
2. Bernstein, S.: Factorization of the nonlinear Schrödinger equation and applications. Complex Var. Elliptic Equ. **51**(5), 429–452 (2006)
3. Bulla, E., Constales, D., Kraußhar, R.S., Ryan, J.: Dirac type operators for arithmetic subgroups of generalized modular groups. Journal für die Reine und Angewandte Mathematik **643**, 1–19 (2010)
4. Cerejeiras, P., Kähler, U.: Elliptic boundary value problems of fluid dynamics over unbounded domains. Math. Methods Appl. Sci. **23**(1), 81–101 (2000)
5. Cerejeiras, P., Kähler, U., Sommen, F.: Parabolic Dirac operators and the Navier-Stokes equations over time-varying domains. Math. Methods Appl. Sci. **28**(14), 1715–1724 (2005)
6. Cerejeiras, P., Vieira, N.: Regularization of the non-stationary Schrödinger operator. Math. Methods Appl. Sci. **32**(4), 535–555 (2009)

7. Constales, D., Kraußhar, R.S.: On the Navier-Stokes equation with Free Convection in three dimensional triangular channels. Math. Methods Appl. Sci. **31**(6), 735–751 (2008)
8. Constales, D., Kraußhar, R.S.: Multiperiodic eigensolutions to the Dirac operator and applications to the generalized Helmholtz equation on flat cylinders and on the n-torus. Math. Methods Appl. Sci. **32**(16), 2050–2070 (2009)
9. Constales, D., Grob, D., Kraußhar, R.S.: On generalized Helmholtz type equations in concentric annular domains in $\mathbb{R}^3$. Math. Meth. Appl. Sci. **33**(4), 431–438 (2010)
10. Delanghe, R., Sommen, S., Souček, V.: Clifford Algebra and Spinor Valued Functions. Kluwer, Dortrecht, Boston, London (1992)
11. Faustino, N., Gürlebeck, K., Hommel, A., Kähler, U.: Difference potentials for the Navier-Stokes equations in unbounded domains. J. Differ. Equ. Appl. **12**(6), 577–595 (2006)
12. Gürlebeck, K., Hommel, A.: On discrete Stokes and Navier-Stokes equations in the plane. In: Ablamowicz, R. (ed.) Clifford algebras. Applications to mathematics, Physics, and Engineering, Progress in Mathematical Physics, vol. 34, pp. 35–58. Birkhäuser, Boston (2004)
13. Gürlebeck, K., Sprößig, S.: Quaternionic Analysis and Elliptic Boundary Value Problems. Birkhäuser, Basel (1990)
14. Gürlebeck, K., Sprößig, S.: Quaternionic and Clifford Calculus for Physicists and Engineers. Wiley, Chichester, New York (1997)
15. Kraußhar, R.S., Ryan, J.: Some conformally flat spin manifolds, dirac operators and automorphic forms. J. Math. Anal. Appl. **325**(1), 359–376 (2007)
16. Kraußhar, R.S., Vieira, N.: The Schrödinger equation on cylinders and the n-torus. J. Evol. Equ. **11**, 215–237 (2011)
17. Kravchenko, V.: Applied Quaternionic Analysis, Research and Exposition in Mathematics, vol. 28. Heldermann Verlag, Lemgo (2003)
18. Kravchenko, V., Shapiro, M.: Integral Representations for Spatial Models of Mathematical Physics. Addison Wesley Longman, Harlow (1996)
19. Kuiper, N.H.: On conformally flat spaces in the large. Ann. Math. **50**, 916–924 (1949)
20. Lawson, H.B., Michelsohn, M.-L.: Spin Geometry. Princeton University Press, New York (1989)
21. Mitrea, M.: Clifford Wavelets, Singular Integrals and Hardy Spaces. Lecture Notes in Mathematics. Springer, New York (1994)
22. Mitrea, M.: Boundary value problems for Dirac operators and Maxwell's equations in nonsmooth domains. Math. Math. Appl. Sci. **25**, 1355–1369 (2002)
23. Sprößig, W.: Forecasting equations in complex quaternionic setting. In: Simos, T.E. (ed.), Recent Advances in Computational and Applied Mathematics, European Academy of Science. Springer, Doordrecht (Chapter 12)
24. Wu, S.: Well-posedness in Sobolev spaces of the full water wave problem in 3D. J. Am. Math. Soc. **12**, 445–495 (1999)

From Hermitean Clifford Analysis to Subelliptic Dirac Operators on Odd Dimensional Spheres and Other CR Manifolds

P. Cerejeiras, U. Kähler and J. Ryan

Abstract We show that the two Dirac operators arising in Hermitian Clifford analysis are identical to standard differential operators arising in several complex variables. We also show that the maximal subgroup that preserves these operators are generated by translations, dilations and actions of the unitary n-group. So the operators are not invariant under Kelvin inversion. We also show that the Dirac operators constructed via two by two matrices in Hermitian Clifford analysis correspond to standard Dirac operators in euclidean space. In order to develop Hermitian Clifford analysis in a different direction we introduce a sub elliptic Dirac operator acting on sections of a bundle over odd dimensional spheres. The particular case of the three sphere is examined in detail. We conclude by indicating how this construction could extend to other CR manifolds.

Keywords Kohn Laplacian · Kohn Dirac operator · CR manifolds

1 Introduction

Clifford analysis started as an attempt to generalize one variable complex analysis to n-dimensional euclidean space. It has since evolved into a study of the analyst,

P. Cerejeiras (✉) · U. Kähler
CIDMA, Departamento de Matemática, Universidade de Aveiro, Campus de Santiago, 3810-193 Aveiro, Portugal
e-mail: pceres@ua.pt

U. Kähler
e-mail: ukaehler@ua.pt

J. Ryan
Department of Mathematical Sciences, University of Arkansas, Fayetteville, AR 72701, USA
e-mail: jryan@uark.edu

© Springer Nature Switzerland AG 2018
P. Cerejeiras et al. (eds.), *Clifford Analysis and Related Topics*,
Springer Proceedings in Mathematics & Statistics 260,
https://doi.org/10.1007/978-3-030-00049-3_3

geometry and applications of Dirac operators over euclidean space, spheres, real projective spaces, conformally flat spin manifolds and spin manifolds with applications to representation theory arising from mathematical physics, classical harmonic analysis and many other topics.

In recent years a topic referred to as Hermitian Clifford analysis has attracted some attention. See for instance [2–4]. It is developed initially over a complexification of even dimensional Euclidean space. An almost complex structure is introduced and two associated projection operators are applied to this complex vector space. This splits this space into two n-dimensional complex spaces. When these operators are applied to the euclidean Dirac operator it splits into two differential operators. Here we show that these operators are respectively the d-bar operator and its dual as arising in several complex variables. See for instance [5, 6].

The Euclidean Dirac operator is a conformally invariant operator. In particular it is invariant under Kelvin inversion. We show that these Hermitian Dirac operators have a narrower range of invariance. We show they are no longer invariant under Kelvin inversion. Their maximal invariance group is generated by translation, dilation and a subgroup of $SO(2n)$ isomorphic to $U(n)$.

We also show via lemma given in [1] that the Dirac operator constructed via two by two matrices is in fact a standard Dirac operator over Euclidean space. In order to try and develop fresh ideas for this topic we transfer to odd dimensional spheres and make use of their CR structure to introduce a subelliptic Dirac operator acting on sections of a bundle defined over the sphere. Each fiber is isomorphic to a Clifford algebra generated from the CR structure of the sphere. The square of this Dirac operator gives the Kohn Laplacian on the sphere. The Hopf vibration is used to illustrate this scenario in more detail over $\mathbb{S}^3$. We conclude by illustrating how this construction might carry over to other CR manifolds.

2 Preliminaries on Hermitean Clifford Analysis

In Hermitian Clifford analysis one starts with the standard Dirac operator

$$D = \sum_{j=1}^{n} e_j \frac{\partial}{\partial x_j} \tag{1}$$

in the Euclidean space $\mathbb{R}^m$. Here

$$e_i e_j + e_j e_i = -2\delta_{ij}, \tag{2}$$

under Clifford algebra multiplication.

So we consider the real 2^m dimensional Clifford algebra $\mathcal{C}\ell_{m,-}$ with $\mathbb{R}^m \subset \mathcal{C}\ell_{m,-}$ and $x^2 = -\|x\|^2$, for all each $x \in \mathbb{R}^m$.

One now introduces an almost complex structure on $\mathbb{R}^m$. This is a matrix $J \in SO(m)$ with $J^2 = -I$. As $J \in SO(m)$ this forces m to be even. So $m = 2n$. For instances, $J = \begin{pmatrix} 0 & -I \\ I & 0 \end{pmatrix}$. This is the choice of J that we will use here. We now complexify $\mathbb{R}^{2n}$ to obtain $\mathbb{C}^{2n}$ and we complexify $\mathscr{C}\ell_{2n,-}$ to obtain the complex Clifford algebra $\mathscr{C}\ell_{2n}(\mathbb{C})$. Following [5, 6] we now have the projection operators $\frac{1}{2}(I \pm i J)$. They act on $\mathbb{C}^{2n}$ and split it into two complex spaces $W^+ \oplus W^-$ each of complex dimension n. So

$$W^\pm = \frac{1}{2}(I \pm i J)\mathbb{C}^{2n}.$$

In particular, $\frac{1}{2}(I \pm i J)e_j = \frac{1}{2}(e_j \pm ie_{j+n})$, for $j = 1, \ldots, n$, and it equals $\frac{1}{2}(e_j \mp ie_{j-n})$, for $j = n+1, \ldots, 2n$. Note $\frac{1}{2}(e_j \mp ie_{j-n}) = \frac{1}{2}i(e_k \pm ie_{k+n})$ for $j = n+1, \ldots, 2n$ and $k = j - n$.

We denote $\frac{1}{2}(e_k \pm ie_{k+n})$ by $\mathfrak{f}_k^\pm$ respectively, with $k = 1, \ldots, n$. The elements $\mathfrak{f}_k^\pm$, $k = 1, \ldots, n$, is known as a Witt basis for $W^\pm$, respectively. Note that $(\mathfrak{f}_k^\pm)^2 = 0$,

$$\mathfrak{f}_j^\pm \mathfrak{f}_k^\pm + \mathfrak{f}_k^\pm \mathfrak{f}_j^\pm = 0,$$

and

$$\mathfrak{f}_j^+ \mathfrak{f}_k^- + \mathfrak{f}_k^- \mathfrak{f}_j^+ = \delta_{j,k}.$$

These relations correspond to the relations of differential forms $dz_j, d\bar{z}_k$, and their duals on the alternating algebra $\wedge(\mathbb{C}^n)$. See, for instance [5, 6]. In fact, $e_j = \mathfrak{f}_j^+ + \mathfrak{f}_j^-$ for $j = 1, \ldots, n$ and $e_j = -i(\mathfrak{f}_j^+ - \mathfrak{f}_j^-)$ for $j = n+1, \ldots, 2n$. So our complex Clifford algebra and the alternating algebra are the same.

Following [2, 4] we now consider

$$\frac{1}{2}(I \pm i J)D.$$

This splits the Dirac operator, D, into a pair of operators $D^\pm$ acting over $W^\pm$ respectively. So $D = D^- \oplus D^+$. Explicitly,

$$D^+ = \frac{1}{2} \begin{pmatrix} D & -iD \\ iD & D \end{pmatrix}, \quad D^- = \frac{1}{2} \begin{pmatrix} D & iD \\ -iD & D \end{pmatrix}.$$

Now consider $\mathbb{C}^n = \{(z_1, \ldots, z_n), z_1, \ldots, z_n \in \mathbb{C}\}$.

We have the mappings

$$P^\pm : \mathbb{C}^n \rightarrow W^\pm, \quad (z_1, \ldots, z_n) \mapsto z_1 \mathfrak{f}_1^\pm + \cdots + z_n \mathfrak{f}_n^\pm.$$

Via these identifications it can be seen that D^+ corresponds to the operator $\overline{\partial} = \sum_{j=1}^{n} d\overline{z}_j \frac{\partial}{\partial \overline{z}_j}$ and D^- corresponds to its dual $\overline{\partial}^*$ from several complex variables. See for instance [5, 6]. In this way the Dirac operator corresponds to the operator $\overline{\partial} + \overline{\partial}^*$.

3 Hermitean Clifford Analysis and Matrix Differential Operators

In [13] the pair of matrix differential operators

$$\begin{pmatrix} D^+ & D^- \\ D^- & D^+ \end{pmatrix}, \quad \begin{pmatrix} D^- & D^+ \\ D^+ & D^- \end{pmatrix}$$

are introduced. Note that

$$\begin{pmatrix} D^+ & D^- \\ D^- & D^+ \end{pmatrix}\begin{pmatrix} D^- & D^+ \\ D^+ & D^- \end{pmatrix} = \begin{pmatrix} D^- & D^+ \\ D^+ & D^- \end{pmatrix}\begin{pmatrix} D^+ & D^- \\ D^- & D^+ \end{pmatrix} = \Delta_{2n}\begin{pmatrix} I & 0 \\ 0 & I \end{pmatrix},$$

where Δ_{2n} is the Laplacian on $\mathbb{R}^{2n}$. The first matrx operator can be written as

$$\begin{pmatrix} D^+ & D^- \\ D^- & D^+ \end{pmatrix} = D^+\begin{pmatrix} I & 0 \\ 0 & I \end{pmatrix} + D^-\begin{pmatrix} 0 & I \\ I & 0 \end{pmatrix}$$

and the second as

$$\begin{pmatrix} D^- & D^+ \\ D^+ & D^- \end{pmatrix} = D^-\begin{pmatrix} I & 0 \\ 0 & I \end{pmatrix} + D^+\begin{pmatrix} 0 & I \\ I & 0 \end{pmatrix}.$$

Now, the space $\mathbb{R}(2)$ of 2×2 matrices is a representation of the Clifford algebra $\mathscr{C}\ell_{2,+}$. See for instances [1]. To see this, consider the basis $1, g_1, g_2, g_1g_2$ of $\mathscr{C}\ell_{2,+}$. So $g_i^2 = 1$ for $i = 1, 2$. Now, make the identification

$$1 \leftrightarrow \begin{pmatrix} I & 0 \\ 0 & I \end{pmatrix}, g_1 \leftrightarrow \begin{pmatrix} I & 0 \\ 0 & -I \end{pmatrix}, g_2 \leftrightarrow \begin{pmatrix} 0 & I \\ I & 0 \end{pmatrix}.$$

So g_1g_2 is identified with $\begin{pmatrix} 0 & -I \\ I & 0 \end{pmatrix}$. This defines an algebra isomorphism between $\mathscr{C}\ell_{2,+}$ and $\mathbb{R}(2)$. The matrix differential operators now become

$$D^+ \otimes 1 + D^- \otimes g_2 \sim D^+ + D^-g_2, \quad D^- \otimes 1 + D^+ \otimes g_2 \sim D^- + D^+g_2,$$

and

$$(D^+ + D^-g_2)(D^- + D^+g_2) = (D^+D^- + D^-D^+) = \Delta_{2n}.$$

Further $\mathscr{C}\ell_{2n,-} \otimes \mathscr{C}\ell_{2,+} \cong \mathscr{C}\ell_{2n+2,+}$ (see [1]). So, $D^+ \otimes 1 + D^- \otimes g_2$ and $D^- \otimes 1 + D^+ \otimes g_2$ define a couple of Dirac operators defined over a copy of $\mathbb{R}^{2n}$ lying in $\mathscr{C}\ell_{2n+2,+}$.

The fundamental solution to $D^+ + D^- g_2 \sim D^+ \otimes 1 + D^- \otimes g_2$ is, up to a constant,

$$E_1(x) = \frac{x^- \otimes 1 + x^+ \otimes g_2}{|x|^{2n}}$$

where $x = x_1 e_1 + \cdots + x_{2n} e_{2n}$ and $x^{\pm} = \frac{1}{2}(I \pm i J)x$.

In matrix form this corresponds to

$$\frac{1}{|x|^{2n}} \begin{pmatrix} x^- & x^+ \\ x^+ & x^- \end{pmatrix},$$

while the fundamental solution to the operator $D^- \otimes 1 + D^+ \otimes g_2$ is, up to a constant,

$$E_2(x) = \frac{x^+ \otimes 1 + x^- \otimes g_2}{|x|^{2n}}.$$

In matrix form this gives

$$\frac{1}{|x|^{2n}} \begin{pmatrix} x^+ & x^- \\ x^- & x^+ \end{pmatrix}.$$

They correspond to the fundamental solutions to the matrix operator given in [3].

The following Borel-Pompeiu and Cauchy integral formulas correspond to the integral formulas appearing in [3].

$$f(y) = \frac{1}{\omega_n} \int_S E_+(x)(n^+(x) \otimes 1 + n^-(x) \otimes g_2) f(x) d\sigma(x)$$

$$+ \frac{1}{\omega_n} \int \int_{\Omega} E_+(x)(D^+ \otimes 1 + D^- \otimes g_2) f(x) d\mu(x)$$

$$= \frac{1}{\omega_n} \int_S E_+(x)(n^+(x) + n^-(x)g_2) f(x) d\sigma(x)$$

$$+ \frac{1}{\omega_n} \int \int_{\Omega} E_+(x)(D^+ + D^- g_2) f(x) d\mu(x),$$

and if $(D^+ \otimes 1 + D^- \otimes g_2) f(x) = 0$ this integral gives a Cauchy integral formula.

There are similar integral formulas obtained by replacing E_+ by E_-, $n^+(x) \otimes 1 + n^-(x) \otimes g_2$ by $n^-(x) \otimes 1 + n^+(x) \otimes g_2$ and $D^+ \otimes 1 + D^- \otimes g_2$ by $D^- \otimes 1 + D^+ \otimes g_2$.

4   Conformal Transformations in the Hermitean Case

In [14] and elsewhere it is shown that the Euclidean Dirac operator is invariant under conformal transformations. A theorem of Liouville tells us that for dimensions 3 and above the only conformal transformations are Möbius transformations *indentified with* $SO(2n + 1, 1)$. These are transformations generated from translations, dilations, orthogonal transformations and the Kelvin inversion $\left(x \rightarrow \frac{x}{|x|^2}\right)$. The question now is what subgroup of the conformal group preserves D^+ or equivalently, $\overline{\partial}$ and $\overline{\partial}^*$? As $D^{\pm}$ are both homogeneous and with constant coefficients they are invariant under both translations and dilations.

As for orthogonal transformations, such a transformation must preserve the spaces $W^{\pm}$. So we need to restrict to transformations in $SO(2n)$ that commute with J. It is known that this is a subgroup of $SO(2n)$ isomorphic to $\mathscr{U}(n)$.

To proceed further we need the spin group. First, consider

$$\{a \in \mathscr{C}\ell_{2n,-} : a = y_1 \cdots y_p, \text{ for } y_1, \ldots, y_p \in \mathbb{S}^{2n-1}\}.$$

This set is a group under Clifford algebra multiplication. It is the pin group and it is denoted by $Pin(2n)$. If we restrict so that p is even we obtain a subgroup called the spin group. It is denoted by $Spin(2n)$.

For $a = y_1 \cdots y_{2p} \in Spin(2n)$ we denote $y_{2p} \cdots y_1$ by $\tilde{a}$.

It is well known, [12] for instance, that for $x \in \mathbb{R}^{2n}$, $ax\tilde{a}$ defines a special orthogonal transformation on $\mathbb{R}^{2n}$. In fact, [12], there is a surjective group homomorphism

$$\theta : Spin(2n) \rightarrow SO(2n), \quad a \mapsto \theta_a,$$

where $\theta_a x := ax\tilde{a}$. The kernel of this homomorphism is $\{\pm 1\}$, [12].

We have previously noted that the operators $D^{\pm}$ are invariant under actions of a copy of $\mathscr{U}(n) \subset SO(2n)$. It follows that this copy of $\mathscr{U}(n)$ has a double covering, $\mathscr{U}'(n) \subset Spin(2n)$, also isomorphic to $\mathscr{U}(n)$. Further, following [8] then if Ω is a domain in $\mathbb{R}^{2n}$ and $y = ax\tilde{a} \in \Omega$, with $a \in \mathscr{U}'(n)$ then if

$$D^{\pm} f(y) = 0 \quad \text{then} \quad D^{\pm}\tilde{a} f(ax\tilde{a}) = 0,$$

where $D^{\pm}$ now acts with respect to the variable x.

This situation differs from what is considered in several variables for the operators $\overline{\partial}$ and $\overline{\partial}^*$ in $\mathbb{C}^n$. There these operators are considered as acting strictly on (p, q) forms, see for instance [5]. The action of $\tilde{a}$ on the function $f(ax\tilde{a})$ is a spherical action that does not preserve (p, q) forms. It preserves the spinor subspaces of the algebra $\mathscr{C}\ell_{2n,-}$ or $\mathscr{C}\ell_{2n}(\mathbb{C})$, see [12].

We now turn to consider the case of the Kelvin inversion. Given a hypersurface S in $\mathbb{R}^{2n}$ then under Kelvin inversion the surface element $n(y)d\sigma(y)$ is transformed to $G(x)n(x)G(x)d\sigma(x)$, where $y \in S$, $n(y)$ is the outer pointing unit vector perpendicular to the tangent space TS_y, and σ is the Lebesgue measure on S. We want to know if the spaces $W^{\pm}$ are preserved under the Kelvin inversion and in particular if the operators $D^{\pm}$ are also preserved. As $G(x) = \frac{x}{|x|^{2n}}$, this boils down to determine whether or not J commutes with the action $xn(x)x$ of x on $n(x)$. Note that $xn(x)x$ describes (up to the sign) a reflection of $n(x)$ in the direction of x. So $xn(x)x$ is a vector in $\mathbb{R}^{2n}$.

As J is a matrix in $SO(2n)$ and the vector $xn(x)x$ is defined in terms of Clifford multiplication it is better to make the notation uniform. The action of J on $\mathbb{R}^{2n}$ gives a counterclockwise rotation of $\pi/2$ in each plane spanned by e_i, e_{i+n} for $i = 1, \dots n$. So a lifting of J to $Spin(2n)$ gives

$$\pm j = \pm \frac{1}{\sqrt{2}}(1 + e_1 e_{1+n}) \cdots \frac{1}{\sqrt{2}}(1 + e_n e_{2n}) = \pm \frac{1}{2^{n/2}}(1 + e_1 e_{1+n}) \cdots (1 + e_n e_{2n}).$$

It should be noted that each pair $(1 + e_i e_{i+n})(1 + e_s e_{s+n})$ commutes with each other. It follows that it is enough to compare the terms

$$(1 + e_1 e_{1+n})xn(x)x(1 + e_{1+n}e_1)$$

and

$$x(1 + e_1 e_{1+n})n(x)(1 + e_{1+n}e_1)x.$$

In fact, it is enough to compare the subterms $(1 + e_1 e_{1+n})xn(x)x$ and $x(1 + e_1 e_{1+n})n(x)x$. As the hypersurface S is arbitrary we can place $x = e_1$. In this case the terms

$$(1 + e_1 e_{1+n})e_1 = (e_1 + e_{1+n}) \quad \text{and} \quad e_1(1 + e_1 e_{1+n}) = (e_1 - e_{1+n})$$

differ by a sign. It follows that the operators $D^{\pm}$ are not invariant under the Kelvin transformation. Consequently the operators $\overline{\partial}$ and $\overline{\partial}^*$ are not invariant under the Kelvin transformation.

Summing up we have established:

Theorem 1 *The operators $D^{\pm}$ (and consequently, $\overline{\partial}$ and $\overline{\partial}^*$) are only invariant under translations, dilations and actions of the group $\mathscr{U}'(n) \subset Spin(2n)$.*

5 Kohn Dirac and Laplacian Operators in S^{2n-1}

In the previous sections we have shown that much of the Hermitean Clifford analysis already exists in several complex variables over $\mathbb{C}^n$. This leads to the question of trying to find ways in which to develop Hermitean Clifford analysis that are more meaningful and original.

One starting point is to look at its analogue over odd dimensional spheres.

Consider $\mathbb{S}^{2n-1} \subset \mathbb{C}^n$. Consider also $x \in \mathbb{S}^{2n-1}$ and $T\mathbb{S}_x^{2n-1} \subset \mathbb{C}^n$. In several complex variables [5, 6] one considers the complex subspace L_x of $T\mathbb{S}_x^{2n-1}$ where

$$L_x = \{y \in T\mathbb{S}_x^{2n-1} : iy \in T\mathbb{S}_x^{2n-1}\}.$$

Here, $n \geq 2$. This gives rise to a situation that does not occur in one complex variable. We now have a complex bundle $\mathbf{L} \subset T\mathbb{S}^{2n-1}$ where each fiber is the complex space L_x. Consider now the canonical projection

$$P_x : T\mathbb{S}_x^{2n-1} \to L_x.$$

This gives rise to the projection

$$\mathbf{P} : T\mathbb{S}_x^{2n-1} \to \mathbf{L}.$$

Now consider a hypersurface S' bounding a domain Ω' in $\mathbb{S}^{2n-1}$. If we consider S' to be sufficiently smooth we can consider smooth $\mathscr{C}\ell_{2n}(\mathbb{C})$−valued functions defined in a neighbourhood of Ω' within $\mathbb{S}^{2n-1}$ and following [10] we have the following version of the Stokes' Theorem.

Theorem 2

$$\int_{S'} f(x)n(x)g(x)d\sigma(x) = \int\int_{\Omega'} ((f(x)D_s)g(x) + f(x)(D_s g(x)))\, dx,$$

where $n(x)$ is the outward pointing unit vector in $T S_x'$. Further D_s is the spherical Dirac operator $x\left(\Gamma_x + \frac{n}{2}\right)$, where $\Gamma_x := \sum_{1 \leq i < j \leq 2n-1} e_i e_j \left(x_i \partial_{x_j} - x_j \partial_{x_i}\right)$.

We may also consider the modified integral

$$\int_{S'} f(x)P_x(n(x))g(x)d\sigma(x).$$

Applying the Stokes' Theorem to this integral gives the integral

$$\int\int_{\Omega'} ((f(x)(D_s P_x))g(x) + f(x)((P_x D_s)g(x)))\, dx.$$

For each $x \in \mathbb{S}^{2n-1}$, we have $ix \in T\mathbb{S}_x^{2n-1}$, so $ix \notin L_x$. Consequently, $P_x(ix) = 0$. Therefore, $P_x(D_s) = P_x(x\Gamma_x)$. It follows that this operator is an isomorphic copy of the sub-elliptic Dirac operator $\overline{\partial}_b + \overline{\partial}_b^*$ over S^{2n-1}. This Dirac operator is also known as the Kohn Dirac operator (see [11]). Therefore, $(P_x(x\Gamma_x))^2 = \Box_b$, the Kohn Laplacian. The operators $\partial_b, \overline{\partial}_b^*$ and $\Box_b$ are all defined in [5, 6] and elsewhere.

The group of conformal transformations that preserves the unit b all $B(0, 1) \subset \mathbb{R}^{2n}$, and its boundary, is $SO(2n, 1)$. However, we have identified $\mathbb{R}^{2n}$ with $\mathbb{C}^n$. We need to restrict to the subgroup of $SO(2n, 1)$ that preserves the Hermitean structure of

$\mathbb{C}^n$. At this point we need to reference or show that this group is $\mathscr{U}(n, 1)$. We also need to show that the invariant group for $P_x(D_s)$ is the double cover of $\mathscr{U}(n, 1)$ within $Spin(2n)$. A start is the following: note that if $(f(x)(D_s P_x)) = (P_x D_s)g(x) = 0$ then we have the following version of the Cauchy's Theorem:

Theorem 3
$$\int_{S'} f(x) P_x(n(x)) g(x) d\sigma(x) = 0.$$

By allowing S' to vary it follows that the group of diffeomorphisms that preserves the Kohn-Dirac operator $P_x(D_s)$ is a subgroup of $Spin(2n, 1)$ whose projection preserves the bundle **L**.

This should be $\mathscr{U}(n, 1)$, with a double cover in $Spin(2n, 1)$ isomorphic to $\mathscr{U}(n, 1)$.

6 Realization on $\mathbb{S}^3$

To make the above points more clear we consider the three-dimensional sphere as a special example. Furthermore, we will consider for the moment quaternionic valued functions.

Let us introduce the following first-order differential operators

$$X_{\mathbf{i}} = x_0 \partial_{x_1} - x_1 \partial_{x_0} + x_3 \partial_{x_2} - x_2 \partial_{x_3},$$
$$X_{\mathbf{j}} = x_0 \partial_{x_2} - x_2 \partial_{x_0} + x_1 \partial_{x_3} - x_3 \partial_{x_1},$$
$$X_{\mathbf{k}} = x_0 \partial_{x_3} - x_3 \partial_{x_0} + x_1 \partial_{x_2} - x_2 \partial_{x_1}.$$

These differential operators are skew-symmetric with respect to the Riemannian surface form dS on $\mathbb{S}^3$ given by

$$dS = \sum_{j=0}^{3} (-1)^j x_j d\hat{x}_j,$$

where $d\hat{x}_j$ is generated from the oriented volume form dx with dx_j being omitted.
The sub-Laplacian $\Box_b$ is given by

$$\Box_b = -X_{\mathbf{i}}^2 - X_{\mathbf{k}}^2$$

while the Laplacian is given by

$$\Delta = -X_{\mathbf{i}}^2 - X_{\mathbf{j}}^2 - X_{\mathbf{k}}^2.$$

Moreover, we have the sub-Dirac operator

$$D_b := iX_i + kX_k.$$

Keep in mind that in general its square is not the sub-Laplacian since $X_j = [X_i, X_k]$.

We can identify $\mathbb{R}^4 \sim \mathbb{H}$ with $\mathbb{C}^2$ in the usual way via $z = z_1 + iz_2$, with $z_1, z_2 \in \mathbb{C}_j := \{a + bj, a, b \in \mathbb{R}\}$. The one-parameter transformation group generated by X_j corresponds to the complex multiplication with $\lambda = (a + bj)$ from the right. The resulting orbits on $\mathbb{S}^3$ span the complex projective space $P^1\mathbb{C}$ and we get the Hopf bundle

$$\pi_R : \mathbb{S}^3 \mapsto P^1\mathbb{C}.$$

The complex line bundle L^l on $P^1\mathbb{C}$ is associated to the character $\chi_l(\lambda) = \lambda^l$.

As usual we can now consider the space Π_N of homogeneous polynomials of degree N in the variables x_0, x_1, x_2, x_3. We denote by H_N the subspace of harmonic polynomials in Π_N. For this subspace we have the classic Fischer decomposition

$$\Pi_N = H_N + \left(\sum_{i=0}^{3} x_i^2\right) \Pi_{N-2}.$$

The space H_N restricted to the sphere $\mathbb{S}^3$ is the eigenspace $\mathbf{H}_N$ of the Laplacian Δ with respect to the eigenvalue $\lambda_N = N(N + 2)$ with dimension $(N + 1)^2$.

Let us now consider the subspace $\Pi_{n,m}, n + m = N$, of Π_N defined by

$$\Pi_{n,m} = \left\{ p \in \Pi_N : p(z_0 e^{it}, z_1 e^{it}, \overline{z_0 e^{it}}, \overline{z_1 e^{it}}) = e^{i(n-m)t} p(z_0, z_1, \overline{z_0}, \overline{z_1}) \right\}$$

and the subspace $H_{n,m} = H_N \cap \Pi_{n,m}$. In this subspace we have $X_j p = (n - m)p$. This leads to the decomposition

$$H_N = \sum_{n+m=N, n,m \geq 0} \oplus H_{n,m}.$$

Furthermore, we have harmonic Fischer decomposition

$$\Pi_{n,m} = H_{n,m} + \left(|w_0|^2 + |w_1|^2\right) \Pi_{n-1,m-1}.$$

Moreover, each of the eigenspaces $H_{n,m}$ can be decomposed in terms of eigenspaces of the Dirac operator,

$$H_{n,m} = M_{n,m} + (w_0 + w_1)M_{n-1,m} + (\overline{w}_0 + \overline{w}_1)M_{n,m-1} + w_0\overline{w}_1 M_{n-1,m-1} + \overline{w}_0 w_1 M_{n-1,m-1},$$

where $M_{k,s} = \{p \in \Pi_{k,s} : Dp = 0\}$.

Furthermore, for any $p \in H_{n,m}$ we have

$$\Delta p = -(X_{\mathbf{i}}^2 + X_{\mathbf{k}}^2)p + (n-m)^2 p = \Box_b p + (n-m)^2 p$$

as well as

$$D_b p = (\mathbf{i}X_{\mathbf{i}} + \mathbf{k}X_{\mathbf{k}})p + \mathbf{j}(n-m)p$$

and we denote the restriction of $H_{n,m}$ to $\mathbb{S}^3$ by $\mathbf{H}_{n,m}$ as well as the restriction of $M_{n,m}$ to $\mathbb{S}^3$ by $\mathbf{M}_{n,m}$.

This leads to the following lemma.

Lemma 1 *The space $\mathbf{M}_{n,m}$ is the eigenspace of the sub-Dirac operator D_b with respect to the eigenvalue*

$$-N - \mathbf{j}l = -(1+\mathbf{j})n - (1-\mathbf{j})m$$

with multiplicity $N + 2$. Here, $l = n - m$, $N = n + m$, $n, m \geq 0$. Furthermore, the space $\mathbf{H}_{n,m}$ is the eigenspace of the sub-Laplacian $\Box_b$ with respect to the eigenvalue

$$N(N+2) - l^2 = 4m^2 + 4m(1 + |l|) + 2|l|$$

with multiplicity $|l| + 2m + 1$.

Let us consider now the subspace F^l of $C^\infty(\mathbb{S}^3)$ of homogeneous sections on the complex line bundle L^l, i.e. $F^l \sim \Gamma(L^l)$ and

$$F^l = \left\{ f \in C^\infty(\mathbb{S}^3) : f(x(a + b\mathbf{j})) = (a + bi)^{-l} f(x), a, b \in \mathbb{R}, a^2 + b^2 = 1 \right\}.$$

The sub-Laplacian can be identified with the horizontal Laplacian

$$\Box_b : \Gamma(L^l) \mapsto \Gamma(L^l)$$

This identification allows us to decompose F^l as

$$F^l \sim \Gamma(L^l) = \sum_{n-m=l, n, m \geq l} \oplus H_{n,m}$$

into eigenspaces of the horizontal Laplacian.

The sub-Dirac operator can be identified with the horizontal Dirac operator

$$D_b : \Gamma(L^l) \mapsto \Gamma(L^l)$$

This identification allows us to decompose F^l as

$$F^l \sim \Gamma(L^l) = \sum_{n-m=l, n, m \geq l} \oplus M_{n,m}$$

into eigenspaces with respect to the horizontal Dirac operator.

7   Subelliptic Dirac Operator

Following [7, 9] it is desirable to ask what manifolds besides $\mathbb{S}^{2n-1}$ admit a subelliptic Dirac operator. Clearly it should be a CR manifold with some sort of spin structure.

In $\mathbb{C}^n$ we saw an invariance under a subgroup of $Spin(2n)$ isomorphic to $\mathcal{U}(n)$. This suggests that we are looking for a CR manifold with principal bundle whose fibers are isomorphic to $\mathcal{U}(n)$ and with a global double cover also with fiber isomorphic to $\mathcal{U}(n)$.

8   Irreducible Representations in $\mathcal{U}'(n)$

Within the Clifford algebras $C\ell_m$ the irreducible representation spaces for $Spin(m)$ are the spinor spaces. Here though we are dealing with the subgroup $\mathcal{U}'(n)$ of $Spin(2n)$.

As point out in [15] the spinor spaces are no longer irreducible subspaces of $\mathcal{U}'(n)$. Furthermore, in several complex variables one is interested specifically in (p, q) sections. If $\underline{z} = a\underline{\omega}\tilde{a}$ and $f(\underline{z})$ is the isomorphic equivalent of a (p, q) form in the context we consider here then in general $\tilde{a}f(a\underline{\omega}\tilde{a})$ is not the equivalent of a (p, q) form.

To overcome this instead of considering $\tilde{a}f(a\underline{\omega}\tilde{a})$ we consider $\tilde{a}f(a\underline{\omega}\tilde{a})a$. Note that first $\tilde{a}f_j^{\pm}a$ remains in $W^{\pm}$, for $j = 1, \ldots, n$ and $a \in \mathcal{U}'(n)$. Note further that $\tilde{a}a = 1$ for each $a \in \mathcal{U}'(n)$, or for that matter $Spin(2n)$.

Consequently given $f_{j_1}^+ \cdots f_{j_r}^+ f_{k_1}^- \cdots f_{k_s}^-$ then

$$\tilde{a}(f_{j_1}^+ \cdots f_{j_r}^+ f_{k_1}^- \cdots f_{k_s}^-)a = \tilde{a}f_{j_1}^+ a\tilde{a}f_{j_2}^+ a \cdots \tilde{a}f_{j_r}^+ a\tilde{a}f_{k_1}^- a \cdots \tilde{a}f_{k_s}^- a.$$

It follows that if $f(\underline{z})$ is the equivalent of a (p, q) form in the context described here then so is $\tilde{a}f(a\underline{\omega}\tilde{a})a$.

Acknowledgements The work of the first and second authors was partially supported by Portuguese funds through the CIDMA—Center for Research and Development in Mathematics and Applications, and the Portuguese Foundation for Science and Technology ("FCT–Fundação para a Ciência e a Tecnologia"), within project UID/MAT/ 0416/2013.

References

1. Atiyah, M.F., Bott, R., Shapiro, A.: Clifford modules. Topology **3**(Suppl. 1), 3–38 (1964)
2. Brackx, F., De Schepper, H., Sommen, F.: The hermitian clifford analysis toolbox. Adv. Appl. Clifford Alg. **18**, 451–487 (2008)
3. Brackx, F., De Knock, B., De Schepper, H., Sommen, F.: On Cauchy and Martinelli-Bochner integral formulae in Hermitian Clifford analysis. Bull. Braz. Math. Soc. (N.S.) **40**(3), 395416 (2009)
4. Brackx, F., De Schepper, H., Eelbode, D., Soucek, V.: The Howe dual pair in Hermitean Clifford analysis. Rev. Mat. Iberoamericana **26**(2), 449–479 (2010)
5. Boggess, A.: CR Manifolds and the Tangential Cauchy Riemann Complex. CRC Press, Studies in Advanced Mathematics (1991)
6. D'Angelo, J.P.: Several Complex Variables and the Geometry of Real Hypersurfaces. CRC Press (1993)
7. Epstein, C.L.: Subelliptic $Spin_{\mathbb{C}}$ Dirac operators I. Ann. Math. **166**, 183–214 (2007)
8. Eastwood, M.G., Ryan, J.: Aspects of Dirac operators in analysis. Milan J. Math. **75**, 91–116 (2007)
9. Kath, I., Ungermann, O.: Spectra of sub-Dirac operators on certain nilmanifolds. Math. Scand., **117**(1) (2015). arXiv:1311.2418
10. Liu, H., Ryan, J.: Clifford analysis techniques for spherical pde. J. Fourier Anal. Appl. **8**(6), 535–564 (2002)
11. Petit, R.: $Spin^c$-structures and Dirac operators on contact manifolds. Differ. Geom. Appl. **22**, 229–252 (2005)
12. Porteous, I.: Clifford Algebras and the Classical Groups. Cambridge University Press (1995)
13. Rocha-Chávez, R., Shapiro, M., Sommen, F.: Integral Theorems for Functions and Differential Forms in $\mathbb{C}^m$, Res. Notes Math., vol. 428, Chapman & Hall/CRC, Boca Raton, FL (2002)
14. Ryan, J.: Conformally covariant operators in Clifford analysis. Z. Anal. Anwend. **14**, 677–704 (1995)
15. Shirrell, S., Walter, R.: Hermitian Clifford Analysis and Its Connections with Representation Theory. submitted

On Some Conformally Invariant Operators in Euclidean Space

C. Ding and J. Ryan

Abstract The aim of this paper is to correct a mistake in earlier work on the conformal invariance of Rarita-Schwinger operators and use the method of correction to develop properties of some conformally invariant operators in the Rarita-Schwinger setting. We also study properties of some other Rarita-Schwinger type operators, for instance, twistor operators and dual twistor operators. This work is also intended as an attempt to motivate the study of Rarita-Schwinger operators via some representation theory. This calls for a review of earlier work by Stein and Weiss.

Keywords Stein-Weiss type operators · Rarita-Schwinger type operators · Almansi-Fischer decomposition · Conformal invariance · Integral formulas

1 Introduction

In representation theory for Lie groups one is interested in irreducible representation spaces. In particular, for the group $SO(m)$ one might consider the representation space of all harmonic functions on $\mathbb{R}^m$. This space is invariant under the action of $O(m)$, but this space is not irreducible. It decomposes into the infinite sum of harmonic polynomials each homogeneous of degree k, $1 < k < \infty$. Each of these spaces is irreducible for $SO(m)$. See for instance [10]. Hence, one may consider functions $f : U \longrightarrow \mathcal{H}_k$ where U is a domain in $\mathbb{R}^m$ and $\mathcal{H}_k$ is the space of real valued harmonic polynomials homogeneous of degree k. If $\mathcal{H}_k$ is the space of Clifford

---

C. Ding (✉) · J. Ryan
Department of Mathematical Sciences, University of Arkansas,
Fayetteville, AR 72701, USA
e-mail: chaoding1985@gmail.com

J. Ryan
e-mail: jryan@uark.edu

© Springer Nature Switzerland AG 2018
P. Cerejeiras et al. (eds.), *Clifford Analysis and Related Topics*,
Springer Proceedings in Mathematics & Statistics 260,
https://doi.org/10.1007/978-3-030-00049-3_4

algebra valued harmonic polynomials homogeneous of degree k, then an Almansi-Fischer decomposition result tells us that

$$\mathscr{H}_k = \mathscr{M}_k \oplus u\mathscr{M}_{k-1}.$$

Here $\mathscr{M}_k$ and $\mathscr{M}_{k-1}$ are spaces of Clifford algebra valued polynomials homogeneous of degree k and $k-1$ in the variable u, respectively and are solutions to the Dirac equation $D_u f(u) = 0$, where D_u is the Euclidean Dirac operator. The elements of these spaces are known as homogeneous *monogenic* polynomials. In this case the underlying group $SO(m)$ is replaced by its double cover $Spin(m)$. See [3].

Classical Clifford analysis is the study of and applications of Dirac type operators. In this case, the functions considered take values in the spinor space, which is an irreducible representation of $Spin(m)$. If we replace the spinor space with some other irreducible representations, for instance, $\mathscr{M}_k$, we will get the Rarita-Schwinger operator as the first generalization of the Dirac operator in higher spin theory. See, for instance [4]. The conformal invariance of this operator, its fundamental solutions and some associated integral formulas were first provided in [4], and then [7]. However, some proofs in [7] rely on the mistake that the Dirac operator in the Rarita-Schwinger setting is also conformally invariant. This will be explained and corrected in Sect. 3.

From the construction of the Rarita-Schwinger operators, we notice that some other Rarita-Schwinger type operators can be constructed similarly, for instance, twistor operators, dual twistor operators and the remaining operators, see [4, 7, 14]. It is worth pointing out that we need to be careful for the reasons we mentioned above when we establish properties for Rarita-Schwinger type operators. Hence, we give the details of proofs of some properties and integral operators for Rarita-Schwinger type operators.

This paper is organized as follows: after a brief introduction to Clifford algebras and Clifford analysis in Sect. 2, representation theory of the Spin group and Stein-Weiss operators are used to motivate Dirac operators and Rarita-Schwinger operators. On the one hand the Dirac operator can be introduced and motivated by an adapted version of Stokes' Theorem. See [9]. Motivation for Rarita-Schwinger operators seem better suited via representation theory, particularly for spin and special orthogonal groups. In Sect. 3, we will use a counter-example to show that the Dirac operator is not conformally invariant in the Rarita-Schwinger setting. Then we give a proof of conformal invariance of the Rarita-Schwinger operators and we provide the intertwining operators for the Rarita-Schwinger operators. Motivated by the Almansi-Fischer decomposition mentioned above, using similar construction with the Rarita-Schwinger operator, we can consider conformally invariant operators between $\mathscr{M}_k$-valued functions and $u\mathscr{M}_{k-1}$-valued functions. This idea brings us other Rarita-Schwinger type operators, for instance, twistor and dual twistor operators. More details of the construction and properties of these operators can be found in Sect. 4.

2 Preliminaries

2.1 Clifford Algebra

A real Clifford algebra, $\mathscr{C}l_m$, can be generated from $\mathbb{R}^m$ by considering the relationship

$$\underline{x}^2 = -\|\underline{x}\|^2$$

for each $\underline{x} \in \mathbb{R}^m$. We have $\mathbb{R}^m \subseteq Cl_m$. If $\{e_1, \ldots, e_m\}$ is an orthonormal basis for $\mathbb{R}^m$, then $\underline{x}^2 = -\|\underline{x}\|^2$ tells us that

$$e_i e_j + e_j e_i = -2\delta_{ij},$$

where δ_{ij} is the Kronecker delta function. Similarly, if we replace $\mathbb{R}^m$ with $\mathbb{C}^m$ in the previous definition and consider the relationship

$$z^2 = -\|z\|^2 = -z_1^2 - z_2^2 - \cdots - z_m^2, \ where \ z = (z_1, z_2, \ldots, z_m) \in \mathbb{C}^m,$$

we get complex Clifford algebra $\mathscr{C}l_m(\mathbb{C})$, which can also be defined as the complexification of the real Clifford algebra

$$\mathscr{C}l_m(\mathbb{C}) = \mathscr{C}l_m \otimes \mathbb{C}.$$

In this paper, we deal with the real Clifford algebra $\mathscr{C}l_m$ unless otherwise specified. An arbitrary element of the basis of the Clifford algebra can be written as $e_A = e_{j_1} \cdots e_{j_r}$, where $A = \{j_1, \ldots, j_r\} \subset \{1, 2, \ldots, m\}$ and $1 \le j_1 < j_2 < \cdots < j_r \le m$. Hence for any element $a \in \mathscr{C}l_m$, we have $a = \sum_A a_A e_A$, where $a_A \in \mathbb{R}$. We will need the following anti-involutions:

- Reversion:

$$\tilde{a} = \sum_A (-1)^{|A|(|A|-1)/2} a_A e_A,$$

where $|A|$ is the cardinality of A. In particular, $\widetilde{e_{j_1} \cdots e_{j_r}} = e_{j_r} \cdots e_{j_1}$. Also $\widetilde{ab} = \tilde{b}\tilde{a}$ for $a, b \in \mathscr{C}l_m$.

- Clifford conjugation:

$$\bar{a} = \sum_A (-1)^{|A|(|A|+1)/2} a_A e_A,$$

satisfying $\overline{e_{j_1} \cdots e_{j_r}} = (-1)^r e_{j_r} \cdots e_{j_1}$ and $\overline{ab} = \bar{b}\bar{a}$ for $a, b \in \mathscr{C}l_m$.

The Pin and Spin groups play an important role in Clifford analysis. The Pin group can be defined as

$$Pin(m) = \{a \in \mathscr{C}l_m : a = y_1 y_2 \dots y_p, \ where \ y_1, \dots, y_p \in \mathbb{S}^{m-1}, \ p \in \mathbb{N}\},$$

where $\mathbb{S}^{m-1}$ is the unit sphere in $\mathbb{R}^m$. $Pin(m)$ is clearly a group under multiplication in $\mathscr{C}l_m$.

Now suppose that $a \in \mathbb{S}^{m-1} \subseteq \mathbb{R}^m$, if we consider axa, we may decompose

$$x = x_{a\parallel} + x_{a\perp},$$

where $x_{a\parallel}$ is the projection of x onto a and $x_{a\perp}$ is the rest, perpendicular to a. Hence $x_{a\parallel}$ is a scalar multiple of a and we have

$$axa = ax_{a\parallel}a + ax_{a\perp}a = -x_{a\parallel} + x_{a\perp}.$$

So the action axa describes a reflection of x across the hyperplane perpendicular to a. By the Cartan-Dieudonné Theorem each $O \in O(m)$ is the composition of a finite number of reflections. If $a = y_1 \cdots y_p \in Pin(m)$, we have $\tilde{a} = y_p \cdots y_1$ and observe that $ax\tilde{a} = O_a(x)$ for some $O_a \in O(m)$. Choosing $y_1, \dots, y_p$ arbitrarily in $\mathbb{S}^{m-1}$, we see that the group homomorphism

$$\theta : Pin(m) \longrightarrow O(m) : a \mapsto O_a, \tag{1}$$

with $a = y_1 \cdots y_p$ and $O_a x = ax\tilde{a}$ is surjective. Further $-ax(-\tilde{a}) = ax\tilde{a}$, so $1, -1 \in Ker(\theta)$. In fact $Ker(\theta) = \{1, -1\}$. See [16]. The Spin group is defined as

$$Spin(m) = \{a \in \mathscr{C}l_m : a = y_1 y_2 \dots y_{2p}, y_1, \dots, y_{2p} \in \mathbb{S}^{m-1}, \ p \in \mathbb{N}\}$$

and it is a subgroup of $Pin(m)$. There is a group homomorphism

$$\theta : Spin(m) \longrightarrow SO(m),$$

which is surjective with kernel $\{1, -1\}$. It is defined by (1). Thus $Spin(m)$ is the double cover of $SO(m)$. See [16] for more details.

For a domain U in $\mathbb{R}^m$, a diffeomorphism $\phi : U \longrightarrow \mathbb{R}^m$ is said to be conformal if, for each $x \in U$ and each $\mathbf{u}, \mathbf{v} \in TU_x$, the angle between $\mathbf{u}$ and $\mathbf{v}$ is preserved under the corresponding differential at x, $d\phi_x$. For $m \geq 3$, a theorem of Liouville tells us the only conformal transformations are Möbius transformations. Ahlfors and Vahlen show that given a Möbius transformation on $\mathbb{R}^m \cup \{\infty\}$ it can be expressed as $y = (ax + b)(cx + d)^{-1}$ where $a, b, c, d \in \mathscr{C}l_m$ and satisfy the following conditions [15]:

1. a, b, c, d are all products of vectors in $\mathbb{R}^m$;
2. $a\tilde{b}$, $c\tilde{d}$, $\tilde{b}c$, $\tilde{d}a \in \mathbb{R}^m$;
3. $a\tilde{d} - b\tilde{c} = \pm 1$.

Since $y = (ax + b)(cx + d)^{-1} = ac^{-1} + (b - ac^{-1}d)(cx + d)^{-1}$, a conformal transformation can be decomposed as compositions of translation, dilation, reflection and inversion. This gives an *Iwasawa decomposition* for Möbius transformations. See [14] for more details. In Sect. 3, we will show that the Rarita-Schwinger operator is conformally invariant.

The Dirac operator in $\mathbb{R}^m$ is defined to be

$$D_x := \sum_{i=1}^{m} e_i \partial_{x_i}.$$

We also let D denote the Dirac operator if there is no confusion in which variable it is with respect to. Note $D_x^2 = -\Delta_x$, where Δ_x is the Laplacian in $\mathbb{R}^m$. A $\mathcal{C}l_m$-valued function $f(x)$ defined on a domain U in $\mathbb{R}^m$ is called left monogenic if $D_x f(x) = 0$. Since multiplication of Clifford numbers is not commutative, there is a similar definition for right monogenic functions.

Let $\mathcal{M}_k$ denote the space of $\mathcal{C}l_m$-valued monogenic polynomials, homogeneous of degree k. Note that if $h_k \in \mathcal{H}_k$, the space of $\mathcal{C}l_m$-valued harmonic polynomials homogeneous of degree k, then $Dh_k \in \mathcal{M}_{k-1}$, but $Dup_{k-1}(u) = (-m - 2k + 2)p_{k-1}u$, so

$$\mathcal{H}_k = \mathcal{M}_k \oplus u\mathcal{M}_{k-1}, \quad h_k = p_k + up_{k-1}.$$

This is an *Almansi-Fischer decomposition* of $\mathcal{H}_k$. See [7] for more details. Similarly, we can obtain by conjugation a right Almansi-Fischer decomposition,

$$\mathcal{H}_k = \overline{\mathcal{M}}_k \oplus \overline{\mathcal{M}}_{k-1}u,$$

where $\overline{\mathcal{M}}_k$ stands for the space of right monogenic polynomials homogeneous of degree k.

In this Almansi-Fischer decomposition, we define P_k as the projection map

$$P_k : \mathcal{H}_k \longrightarrow \mathcal{M}_k.$$

Suppose U is a domain in $\mathbb{R}^m$. Consider $f : U \times \mathbb{R}^m \longrightarrow \mathcal{C}l_m$, such that for each $x \in U$, $f(x, u)$ is a left monogenic polynomial homogeneous of degree k in u, then the Rarita-Schwinger operator is defined as follows

$$R_k := P_k D_x f(x, u) = (\frac{u D_u}{m + 2k - 2} + 1) D_x f(x, u).$$

We also have a right projection $P_{k,r} : \mathscr{H}_k \longrightarrow \overline{\mathscr{M}}_k$, and a right Rarita-Schwinger operator $R_{k,r} = D_x P_{k,r}$. See [4, 7].

2.2 Irreducible Representations of the Spin Group

To motivate the Rarita-Schwinger operators and to be relatively self-contained we cover in the rest of Sect. 2 some basics on representation theory.

Definition 1 A Lie group is a smooth manifold G which is also a group such that multiplication $(g, h) \mapsto gh : G \times G \longrightarrow G$ and inversion $g \mapsto g^{-1} : G \longrightarrow G$ are both smooth.

Let G be a Lie group and V a vector space over $\mathbb{F}$, where $\mathbb{F} = \mathbb{R}$ or $\mathbb{C}$. A *representation* of G is a pair (V, τ) in which τ is a homomorphism from G into the group $Aut(V)$ of invertible $\mathbb{F}$-linear transformations on V. Thus $\tau(g)$ and its inverse $\tau(g)^{-1}$ are both $\mathbb{F}$-linear operators on V such that

$$\tau(g_1 g_2) = \tau(g_1)\tau(g_2), \quad \tau(g^{-1}) = \tau(g)^{-1}$$

for all g_1, g_2 and g in G. In practice, it will often be convenient to think and speak of V as simply a *G-module*. A subspace U in V which is *G-invariant* in the sense that $gu \in U$ for all $g \in G$ and $u \in U$, is called a *submodule* of V or a *subrepresentation*. The dimension of V is called the dimension of the representation. If V is finite-dimensional it is said to be *irreducible* when it contains no submodules other than 0 and itself; otherwise, it is said to be *reducible*. The following three representation spaces of the Spin group are frequently used in Clifford analysis.

2.2.1 Spinor Representation Space $\mathscr{S}$

The most commonly used representation of the Spin group in $\mathscr{C}l_m(\mathbb{C})$ valued function theory is the spinor space. The construction is as follows:

Let us consider complex Clifford algebra $\mathscr{C}l_m(\mathbb{C})$ with even dimension $m = 2n$. $\mathbb{C}^m$ or the space of vectors is embedded in $\mathscr{C}l_m(\mathbb{C})$ as

$$(x_1, x_2, \ldots, x_m) \mapsto \sum_{j=1}^{m} x_j e_j : \mathbb{C}^m \hookrightarrow \mathscr{C}l_m(\mathbb{C}).$$

Define the *Witt basis* elements of $\mathbb{C}^{2n}$ as

$$f_j := \frac{e_j - i e_{j+n}}{2}, \quad f_j^\dagger := -\frac{e_j + i e_{j+n}}{2}.$$

Let $I := f_1 f_1^\dagger \dots f_n f_n^\dagger$. The space of *Dirac spinors* is defined as

$$\mathscr{S} := \mathcal{C}l_m(\mathbb{C})I.$$

This is a representation of $Spin(m)$ under the following action

$$\rho(s)I := sI, \quad for\ s \in Spin(m).$$

Note that $\mathscr{S}$ is a left ideal of $\mathcal{C}l_m(\mathbb{C})$. For more details, we refer the reader to [6]. An alternative construction of spinor spaces is given in the classical paper of Atiyah, Bott and Shapiro [1].

2.2.2 Homogeneous Harmonic Polynomials on $\mathscr{H}_k(\mathbb{R}^m, \mathbb{C})$

It is a well-known fact that the space of harmonic polynomials is invariant under the action of $Spin(m)$, since the Laplacian Δ_m is an $SO(m)$ invariant operator. But it is not irreducible for $Spin(m)$. It can be decomposed into the infinite sum of k-homogeneous harmonic polynomials, $1 < k < \infty$. Each of these spaces is irreducible for $Spin(m)$. This brings us the most familiar representations of $Spin(m)$: spaces of k-homogeneous harmonic polynomials on $\mathbb{R}^m$. The following action has been shown to be an irreducible representation of $Spin(m)$ (see [13]):

$$\rho : Spin(m) \longrightarrow Aut(\mathscr{H}_k), \quad s \longmapsto \left(f(x) \mapsto \tilde{s} f(sx\tilde{s})s\right)$$

This can also be realized as follows

$$Spin(m) \xrightarrow{\theta} SO(m) \xrightarrow{\rho} Aut(\mathscr{H}_k);$$
$$a \longmapsto O_a \longmapsto \left(f(x) \mapsto f(O_a x)\right),$$

where θ is the double covering map and ρ is the standard action of $SO(m)$ on a function $f(x) \in \mathscr{H}_k$ with $x \in \mathbb{R}^m$.

2.2.3 Homogeneous Monogenic Polynomials on $\mathcal{C}l_m$

In $\mathcal{C}l_m$-valued function theory, the previously mentioned Almansi-Fischer decomposition shows us we can also decompose the space of k-homogeneous harmonic polynomials as follows

$$\mathscr{H}_k = \mathcal{M}_k \oplus u \mathcal{M}_{k-1}.$$

If we restrict $\mathcal{M}_k$ to the spinor valued subspace, we have another important representation of $Spin(m)$: the space of k-homogeneous spinor-valued monogenic polynomials on $\mathbb{R}^m$, henceforth denoted by $\mathscr{M}_k := \mathscr{M}_k(\mathbb{R}^m, \mathscr{S})$. More specifically, the following action has been shown as an irreducible representation of $Spin(m)$:

$$\pi : Spin(m) \longrightarrow Aut(\mathscr{M}_k), \ s \longmapsto f(x) \mapsto \tilde{s} f(s x \tilde{s}).$$

For more details, we refer the reader to [17].

2.2.4 Stein-Weiss Operators

Let U and V be m-dimensional inner product vector spaces over a field $\mathbb{F}$. Denote the groups of all automorphism of U and V by $GL(U)$ and $GL(V)$, respectively. Suppose $\rho_1 : G \longrightarrow GL(U)$ and $\rho_2 : G \longrightarrow GL(V)$ are irreducible representations of a compact Lie group G. We have a function $f : U \longrightarrow V$ which has continuous derivative. Taking the gradient of the function $f(x)$, we have

$$\nabla f \in Hom(U, V) \cong U^* \otimes V \cong U \otimes V, \ where \ \nabla := (\partial_{x_1}, \dots, \partial_{x_m}).$$

Denote by $U[\times]V$ the irreducible representation of $U \otimes V$ whose representation space has largest dimension [11]. This is known as the Cartan product of ρ_1 and ρ_2 [8]. Using the inner products on U and V, we may write

$$U \otimes V = (U[\times]V) \oplus (U[\times]V)^{\perp}$$

If we denote by E and $E^{\perp}$ the orthogonal projections onto $U[\times]V$ and $(U[\times]V)^{\perp}$, respectively, then we define differential operators D and $D^{\perp}$ associated to ρ_1 and ρ_2 by

$$D = E\nabla; \ D^{\perp} = E^{\perp}\nabla.$$

These are called *Stein-Weiss type operators* after [21]. The importance of this construction is that you can reconstruct many first order differential operators with it when you choose proper representation spaces U and V for a Lie group G. For instance, Euclidean Dirac operators [20, 21] and Rarita-Schwinger operators [10]. The connections are as follows:

1. Dirac operators
Here we only show the odd dimension case. Similar arguments also apply in the even dimensional case.

Theorem 1 *Let ρ_1 be the representation of the spin group given by the standard representation of $SO(m)$ on $\mathbb{R}^m$*

$$\rho_1 : Spin(m) \longrightarrow SO(m) \longrightarrow GL(\mathbb{R}^m)$$

and let ρ_2 be the spin representation on the spinor space $\mathscr{S}$. Then the Euclidean Dirac operator is the differential operator given by $\mathbb{R}^m[\times]\mathscr{S}$ when $m = 2n + 1$.

Outline Proof: Let $\{e_1, \ldots, e_m\}$ be the orthonormal basis of $\mathbb{R}^m$ and $x = (x_1, \ldots, x_m)$ $\in \mathbb{R}^m$. For a function $f(x)$ having values in $\mathscr{S}$, we must show that the system

$$\sum_{i=1}^{m} e_i \frac{\partial f}{\partial x_i} = 0$$

is equivalent to the system

$$D^{\perp} f = E^{\perp} \nabla f = 0.$$

Since we have

$$\mathbb{R}^m \otimes \mathscr{S} = \mathbb{R}^m[\times]\mathscr{S} \oplus (\mathbb{R}^m[\times]\mathscr{S})^{\perp}$$

and [21] provides us an embedding map

$$\eta : \mathscr{S} \hookrightarrow \mathbb{R}^m \otimes \mathscr{S},$$

$$\omega \mapsto \frac{1}{\sqrt{m}}(e_1\omega, \ldots, e_m\omega).$$

Actually, this is an isomorphism from $\mathscr{S}$ into $\mathbb{R}^m \otimes \mathscr{S}$. For the proof, we refer the reader to *page 175* of [21]. Thus, we have

$$\mathbb{R}^m \otimes \mathscr{S} = \mathbb{R}^m[\times]\mathscr{S} \oplus \eta(\mathscr{S}).$$

Consider the equation $D^{\perp} f = E^{\perp} \nabla f = 0$, where f has values in $\mathscr{S}$. So ∇f has values in $\mathbb{R}^m \otimes \mathscr{S}$, and so the condition $D^{\perp} f = 0$ is equivalent to ∇f being orthogonal to $\eta(\mathscr{S})$. This is precisely the statement that

$$\sum_{i=1}^{m} (\frac{\partial f}{\partial x_i}, e_i\omega) = 0, \ \forall \omega \in \mathscr{S}.$$

Notice, however, that as an endomorphism of $\mathbb{R}^m \otimes \mathscr{S}$, we have $-e_i$ as the dual of e_i, hence the equation above becomes

$$\sum_{i=1}^{m} (e_i \frac{\partial f}{\partial x_i}, \omega) = 0, \ \forall \omega \in \mathscr{S},$$

which says precisely that f must be in the kernel of the Euclidean Dirac operator. This completes the proof. $\qquad \square$

2. Rarita-Schwinger operators

Theorem 2 *Let ρ_1 be defined as above and ρ_2 is the representation of $Spin(m)$ on $\mathscr{M}_k$. Then as a representation of $Spin(m)$, we have the following decomposition*

$$\mathscr{M}_k \otimes \mathbb{R}^m \cong \mathscr{M}_k[\times]\mathbb{R}^m \oplus \mathscr{M}_k \oplus \mathscr{M}_{k-1} \oplus \mathscr{M}_{k,1},$$

where $\mathcal{M}_{k,1}$ is a simplicial monogenic polynomial space as a $Spin(m)$ representation (see more details in [2]). The Rarita-Schwinger operator is the differential operator given by projecting the gradient onto the $\mathcal{M}_k$ component.

Proof Consider $f(x, u) \in C^{\infty}(\mathbb{R}^m, \mathcal{M}_k)$. We observe that the gradient of $f(x, u)$ satisfies

$$\nabla f(x, u) = (\partial_{x_1}, \ldots, \partial_{x_m}) f(x, u) = (\partial_{x_1} f(x, u), \ldots, \partial_{x_m} f(x, u)) \in \mathcal{M}_k \otimes \mathbb{R}^m.$$

A similar argument as in *page 181* of [21] shows

$$\mathcal{M}_k \otimes \mathbb{R}^m = \mathcal{M}_k[\times]\mathbb{R}^m \oplus V_1 \oplus V_2 \oplus V_3,$$

where $V_1 \cong \mathcal{M}_k$, $V_2 \cong \mathcal{M}_{k-1}$ and $V_3 \cong \mathcal{M}_{k,1}$ as $Spin(m)$ representations. Similar arguments as on *page 175* of [21] show

$$\theta: \ \mathcal{M}_k \longrightarrow \mathcal{M}_k \otimes \mathbb{R}^m, \ q_k(u) \mapsto (q_k(u)e_1, \ldots, q_k(u)e_m)$$

is an isomorphism from $\mathcal{M}_k$ into $\mathcal{M}_k \otimes \mathbb{R}^m$. Hence, we have

$$\mathcal{M}_k \otimes \mathbb{R}^m = \mathcal{M}_k[\times]\mathbb{R}^m \oplus \theta(\mathcal{M}_k) \oplus V_2 \oplus V_3.$$

Let P_k' be the projection map from $\mathcal{M}_k \otimes \mathbb{R}^m$ to $\theta(\mathcal{M}_k)$. Consider the equation $P_k'\nabla f(x, u) = 0$ for $f(x, u) \in C^{\infty}(\mathbb{R}^m, \mathcal{M}_k)$. Then, for each fixed x, $\nabla f(x, u) \in \mathcal{M}_k \otimes \mathbb{R}^m$ and the condition $P_k'\nabla f(x, u) = 0$ is equivalent to ∇f being orthogonal to $\theta(\mathcal{M}_k)$. This says precisely

$$\sum_{i=1}^m (q_k(u)e_i, \partial_{x_i} f(x, u))_u = 0, \ \forall q_k(u) \in \mathcal{M}_k,$$

where $(p(u), q(u))_u = \int_{\mathbb{S}^{m-1}} \overline{p(u)} q(u) dS(u)$ is the Fischer inner product for any pair of $\mathcal{C}l_m$-valued polynomials. Since $-e_i$ is the dual of e_i as an endomorphism of $\mathcal{M}_k \otimes \mathbb{R}^m$, the previous equation becomes

$$\sum_{i=1}^m (q_k(u), e_i \partial_{x_i} f(x, u)) = (q_k(u), D_x f(x, u))_u = 0.$$

Since $f(x, u) \in \mathcal{M}_k$ for fixed x, then $D_x f(x, u) \in \mathcal{H}_k$. According to the Almansi-Fischer decomposition, we have

$$D_x f(x, u) = f_1(x, u) + u f_2(x, u), \ f_1(x, u) \in \mathcal{M}_k \text{ and } f_2(x, u) \in \mathcal{M}_{k-1}.$$

We then obtain $(q_k(u), f_1(x, u))_u + (q_k(u), uf_2(x, u))_u = 0$. However, the Clifford-Cauchy theorem [7] shows $(q_k(u), uf_2(x, u))_u = 0$. Thus, the equation $P'_k \nabla f(x, u) = 0$ is equivalent to

$$(q_k(u), f_1(x, u))_u = 0, \ \forall q_k(u) \in \mathcal{M}_k.$$

Hence, $f_1(x, u) = 0$. We also know, from the construction of the Rarita-Schwinger operator, that $f_1(x, u) = R_k f(x, u)$. Therefore, the Stein-Weiss type operator $P'_k \nabla$ is precisely the Rarita-Schwinger operator in this context.

3 Properties of the Rarita-Schwinger Operator

3.1 A Counterexample

We know that the Dirac operator D_x is conformally invariant in $\mathcal{C}l_m$-valued function theory [19]. But in the Rarita-Schwinger setting, D_x is not conformally invariant anymore. In other words, in $\mathcal{C}l_m$-valued function theory, the Dirac operator D_x has the following conformal invariance property under inversion: If $D_x f(x) = 0$, $f(x)$ is a $\mathcal{C}l_m$-valued function and $x = y^{-1}, x \in \mathbb{R}^m$, then $D_y \dfrac{y}{\|y\|^m} f(y^{-1}) = 0$. In the Rarita-Schwinger setting, if $D_x f(x, u) = D_u f(x, u) = 0$, $f(x, u)$ is a polynomial for any fixed $x \subset \mathbb{R}^m$ and let $x = y^{-1}$, $u = \dfrac{y w y}{\|y\|^2}, x \in \mathbb{R}^m$, then $D_y \dfrac{y}{\|y\|^m} f(y^{-1}, \dfrac{y w y}{\|y\|^2}) \neq 0$ in general.

A quick way to see this is to choose the function $f(x, u) = u_1 e_1 - u_2 e_2$, and use $u = \dfrac{y w y}{\|y\|^2} = w - 2 \dfrac{y}{\|y\|^2} \langle w, y \rangle, u_i = w_i - 2 \dfrac{y_i}{\|y\|^2} \langle w, y \rangle,$ where $i = 1, 2, \ldots, m$. A straightforward calculation shows that

$$D_y \frac{y}{\|y\|^m} f(y^{-1}, \frac{y w y}{\|y\|^2}) = \frac{-2wy(y_1 e_1 - y_2 e_2)}{\|y\|^{m+2}} \neq 0,$$

for $m > 2$. However, $P_1 D_y \dfrac{y}{\|y\|^m} f(y^{-1}, \dfrac{y w y}{\|y\|^2}) = \left(\dfrac{w D_w}{m} + 1 \right) w \dfrac{-2y(y_1 e_1 - y_2 e_2)}{\|y\|^{m+2}}$
$= 0$.

3.2 Conformal Invariance

In [7], the conformal invariance of the equation $R_k f = 0$ is proved and some other properties under the assumption that D_x is still conformally invariant in the Rarita-Schwinger setting. This is incorrect as we just showed. In this section, we will use the Iwasawa decomposition of Möbius transformations and some integral formulas to correct this. As observed earlier, according to this Iwasawa decomposition, a conformal transformation is a composition of translation, dilation, reflection and inversion. A simple observation shows that the Rarita-Schwinger operator is conformally invariant under translation and dilation and the conformal invariance under reflection can be found in [13]. Hence, we only show it is conformally invariant under inversion here.

Theorem 3 *For any fixed $x \in U \subset \mathbb{R}^m$, let $f(x, u)$ be a left monogenic polynomial homogeneous of degree k in u. If $R_{k,u} f(x, u) = 0$, then $R_{k,w} G(y) f(y^{-1}, \dfrac{ywy}{\|y\|^2}) = 0$, where $G(y) = \dfrac{y}{\|y\|^m}$, $x = y^{-1}$, $u = \dfrac{ywy}{\|y\|^2} \in \mathbb{R}^m$.*

To establish the conformal invariance of R_k, we need *Stokes' Theorem* for R_k.

Theorem 4 ([7], Stokes' Theorem for R_k) *Let Ω' and Ω be domains in $\mathbb{R}^m$ and suppose the closure of Ω lies in Ω'. Further suppose the closure of Ω is compact and $\partial\Omega$ is piecewise smooth. Let $f, g \in C^1(\Omega', \mathcal{M}_k)$. Then*

$$
\int_\Omega \left[(g(x, u) R_k, f(x, u))_u + (g(x, u), R_k f(x, u)) \right] dx^m
$$
$$
= \int_{\partial\Omega} (g(x, u), P_k d\sigma_x f(x, u))_u
$$
$$
= \int_{\partial\Omega} (g(x, u) d\sigma_x P_{k,r}, f(x, u))_u,
$$

where P_k and $P_{k,r}$ are the left and right projections, $d\sigma_x = n(x) d\sigma(x)$, $d\sigma(x)$ is the area element. $(P(u), Q(u))_u = \int_{\mathbb{S}^{m-1}} P(u) Q(u) dS(u)$ is the inner product for any pair of $\mathcal{C}l_m$-valued polynomials.

If both $f(x, u)$ and $g(x, u)$ are solutions of R_k, then we have *Cauchy's theorem*.

Corollary 1 ([7], Cauchy's Theorem for R_k) *If $R_k f(x, u) = 0$ and $g(x, u) R_k = 0$ for $f, g \in C^1(, \Omega', \mathcal{M}_k)$, then*

$$
\int_{\partial\Omega} (g(x, u), P_k d\sigma_x f(x, u))_u = 0.
$$

We also need the following well-known result.

Proposition 1 ([18]) *Suppose that S is a smooth, orientable surface in R^m and f, g are integrable $\mathcal{C}l_m$-valued functions. Then if $M(x)$ is a conformal transformation, we have*

$$
\int_S f(M(x)) n(M(x)) g(M(x)) ds = \int_{M^{-1}(S)} f(M(x)) \tilde{J}_1(M, x) n(x) J_1(M, x) g(M(x)) dM^{-1}(S),
$$

where $M(x) = (ax + b)(cx + d)^{-1}$, $M^{-1}(S) = \{x \in \mathbb{R}^m : M(x) \in S\}$, $J_1(M, x) = \dfrac{cx + d}{\|cx + d\|^m}$.

Now we are ready to prove *Theorem* 3.

Proof First, in Cauchy's theorem, we let $g(x, u)R_{k,r} = R_k f(x, u) = 0$. Then we have

$$0 = \int_{\partial\Omega} \int_{\mathbb{S}^{m-1}} g(x, u) P_k n(x) f(x, u) dS(u) d\sigma(x)$$

Let $x = y^{-1}$, according to *Proposition* 1, we have

$$= \int_{\partial\Omega^{-1}} \int_{\mathbb{S}^{m-1}} g(u) P_{k,u} G(y) n(y) G(y) f(y^{-1}, u) dS(u) d\sigma(y),$$

where $G(y) = \dfrac{y}{\|y\|^m}$. Set $u = \dfrac{ywy}{\|y\|^2}$, since $P_{k,u}$ interchanges with $G(y)$ [14], we have

$$= \int_{\partial\Omega^{-1}} \int_{\mathbb{S}^{m-1}} g(\frac{ywy}{\|y\|^2}) G(y) P_{k,w} n(y) G(y) f(y^{-1}, \frac{ywy}{\|y\|^2}) dS(w) d\sigma(y)$$

$$= \int_{\partial\Omega^{-1}} (g(\frac{ywy}{\|y\|^2}) G(y), P_{k,w} d\sigma_y G(y) f(y^{-1}, \frac{ywy}{\|y\|^2}))_w,$$

According to Stokes' theorem,

$$= \int_{\Omega^{-1}} (g(\frac{ywy}{\|y\|^2}) G(y), R_{k,w} G(y) f(y^{-1}, \frac{ywy}{\|y\|^2}))_w$$

$$+ \int_{\Omega^{-1}} (g(\frac{ywy}{\|y\|^2}) G(y) R_{k,w}, G(y) f(y^{-1}, \frac{ywy}{\|y\|^2}))_w.$$

Since $g(x, u)$ is arbitrary in the kernel of $R_{k,r}$ and $f(x, u)$ is arbitrary in the kernel of R_k, we get $g(\dfrac{ywy}{\|y\|^2}) G(y) R_{k,w} = R_{k,w} G(y) f(y^{-1}, \dfrac{ywy}{\|y\|^2}) = 0$.

3.3   Intertwining Operators of R_k

In $\mathscr{C}l_m$-valued function theory, if we have the Möbius transformation $y = \phi(x) = (ax + b)(cx + d)^{-1}$ and D_x is the Dirac operator with respect to x and D_y is the Dirac operator with respect to y then $D_x = J_{-1}^{-1}(\phi, x) D_y J_1(\phi, x)$, where $J_{-1}(\phi, x) = \dfrac{cx + d}{\|cx + d\|^{m+2}}$ and $J_1(\phi, x) = \dfrac{cx + d}{\|cx + d\|^m}$ [18]. In the Rarita-Schwinger setting, we have a similar result:

Theorem 5 ([7]) *For any fixed $x \in U \subset \mathbb{R}^m$, let $f(x, u)$ be a left monogenic polynomial homogeneous of degree k in u. Then*

$$J_{-1}^{-1}(\phi, y) R_{k,y,\omega} J_1(\phi, y) f(\phi(y), \frac{\widetilde{(cy + d)}\omega(cy + d)}{\|cy + d\|^2}) = R_{k,x,u} f(x, u),$$

where $x = \phi(y) = (ay + b)(cy + d)^{-1}$ is a Möbius transformation., $u = \dfrac{\widetilde{(cy + d)}\omega(cy + d)}{\|cy + d\|^2}$, $R_{k,x,u}$ and $R_{k,y,\omega}$ are Rarita-Schwinger operators.

Proof We use the techniques in [9] to prove this Theorem. Let $f(x, u)$, $g(x, u) \in C^\infty(\Omega', \mathscr{C}l_m)$ and Ω and Ω' are as in Theorem 4. We have

$$\int_{\partial\Omega} (g(x, u), P_k n(x) f(x, u))_u dx^m$$

$$= \int_{\phi^{-1}(\partial\Omega)} \left(g(\phi(y), \frac{y\omega y}{\|y\|^2}) P_k J_1(\phi, y) n(y) J_1(\phi, y) f(\phi(y), \frac{y\omega y}{\|y\|^2}) \right)_\omega dy^m$$

$$= \int_{\phi^{-1}(\partial\Omega)} \left(g(\phi(y), \frac{y\omega y}{\|y\|^2}) J_1(\phi, y), P_k n(y) J_1(\phi, y) f(\phi(y), \frac{y\omega y}{\|y\|^2}) \right)_\omega dy^m$$

Then we apply the Stokes' Theorem for R_k,

$$\int_{\phi^{-1}(\Omega)} \left(g(\phi(y), \frac{y\omega y}{\|y\|^2}) J_1(\phi, y) R_k, J_1(\phi, y) f(\phi(y), \frac{y\omega y}{\|y\|^2}) \right)_\omega$$

$$+ \left(g(\phi(y), \frac{y\omega y}{\|y\|^2}) J_1(\phi, y), R_k J_1(\phi, y) f(\phi(y), \frac{y\omega y}{\|y\|^2}) \right)_\omega dy^m, \qquad (2)$$

where $u = \dfrac{y\omega y}{\|y\|^2}$. On the other hand,

$$\int_{\partial\Omega} (g(x, u), P_k n(x) f(x, u))_u dx^m$$

$$= \int_\Omega \left[(g(x, u) R_k, f(x, u))_u + (g(x, u), R_k f(x, u))_u \right] dx^m$$

$$= \int_{\phi^{-1}(\Omega)} \left[(g(x, u) R_k, f(x, u))_u + (g(x, u), R_k f(x, u))_u \right] j(y) dy^m$$

$$= \int_{\phi^{-1}(\Omega)} \left[(g(x, u) R_k, f(x, u) j(y))_u + (g(x, u), J_1(\phi, y) J_{-1}(\phi, y) R_k f(x, u))_u \right] dy^m, \qquad (3)$$

where $j(y) = J_{-1}(\phi, y) J_1(\phi, y)$ is the Jacobian. Now, we let arbitrary $g(x, u) \in \ker R_{k,r}$ and since $J_1(\phi, y) g(\phi(y), \frac{y\omega y}{\|y\|^2}) R_{k,r} = 0$, then from (2) and (3), we get

$$\int_{\phi^{-1}(\Omega)} \left(g(\phi(y), \frac{y\omega y}{\|y\|^2}) J_1(\phi, y) R_k J_1(\phi, y) f(\phi(y), \frac{y\omega y}{\|y\|^2})\right)_\omega dy^m$$

$$= \int_{\phi^{-1}(\Omega)} \left(g(\phi(y), \frac{y\omega y}{\|y\|^2}), J_1(\phi, y) J_{-1}(\phi, y) R_k f(x, u)\right)_u dy^m$$

$$= \int_{\phi^{-1}(\Omega)} \left(g(\phi(y), \frac{y\omega y}{\|y\|^2}) J_1(\phi, y) J_{-1}(\phi, y) R_k f(x, u)\right)_\omega dy^m$$

Since Ω is an arbitrary domain in $\mathbb{R}^m$, we have

$$\left(g(\phi(y), \frac{y\omega y}{\|y\|^2}) J_1(\phi, y) R_k J_1(\phi, y) f(\phi(y), \frac{y\omega y}{\|y\|^2})\right)_\omega = \left(g(\phi(y), \frac{y\omega y}{\|y\|^2}) J_1(\phi, y) J_{-1}(\phi, y) R_k f(x, u)\right)_\omega$$

Also, $g(x, u)$ is arbitrary, we get

$$J_1(\phi, y) R_k J_1(\phi, y) f(\phi(y), \frac{y\omega y}{\|y\|^2}) = J_1(\phi, y) J_{-1}(\phi, y) R_k f(x, u).$$

Theorem 5 follows immediately.

4 Rarita-Schwinger Type Operators

In the construction of the Rarita-Schwinger operator above, we notice that the Rarita-Schwinger operator is actually a projection map P_k followed by the Dirac operator D_x, where in the Almansi-Fischer decomposition,

$$\mathscr{M}_k \xrightarrow{D_x} \mathscr{H}_k \otimes \mathscr{S} = \mathscr{M}_k \oplus u\mathscr{M}_{k-1}$$
$$P_k : \mathscr{H}_k \otimes \mathscr{S} \longrightarrow \mathscr{M}_k;$$
$$I - P_k : \mathscr{H}_k \otimes \mathscr{S} \longrightarrow u\mathscr{M}_{k-1}.$$

If we project to the $u\mathscr{M}_{k-1}$ component after we apply D_x, we get a Rarita-Schwinger type operator from $\mathscr{M}_k$ to $u\mathscr{M}_{k-1}$.

$$\mathscr{M}_k \xrightarrow{D_x} \mathscr{H}_k \otimes \mathscr{S} \xrightarrow{I-P_k} u\mathscr{M}_{k-1}.$$

Similarly, starting with $u\mathscr{M}_{k-1}$, we get another two Rarita-Schwinger type operators.

$$u\mathscr{M}_{k-1} \xrightarrow{D_x} \mathscr{H}_k \otimes \mathscr{S} \xrightarrow{P_k} \mathscr{M}_k;$$
$$u\mathscr{M}_{k-1} \xrightarrow{D_x} \mathscr{H}_k \otimes \mathscr{S} \xrightarrow{I-P_k} u\mathscr{M}_{k-1}.$$

In a summary, there are three further Rarita-Schwinger type operators as follows:

$$T_k^* : C^\infty(\mathbb{R}^m, \mathcal{M}_k) \longrightarrow C^\infty(\mathbb{R}^m, u\mathcal{M}_{k-1}), \quad T_k^* = (I - P_k)D_x = \frac{-uD_u}{m + 2k - 2}D_x;$$

$$T_k : C^\infty(\mathbb{R}^m, u\mathcal{M}_{k-1}) \longrightarrow C^\infty(\mathbb{R}^m, \mathcal{M}_k), \quad T_k = P_k D_x = (\frac{uD_u}{m + 2k - 2} + 1)D_x;$$

$$Q_k : C^\infty(\mathbb{R}^m, u\mathcal{M}_{k-1}) \longrightarrow C^\infty(\mathbb{R}^m, u\mathcal{M}_{k-1}), \quad Q_k = (I - P_k)D_x = \frac{-uD_u}{m + 2k - 2}D_x,$$

T_k^* and T_k are also called the *dual-twistor operator* and *twistor operator*. See [4]. We also have

$$T_{k,r}^* : C^\infty(\mathbb{R}^m, \overline{\mathcal{M}}_k) \longrightarrow C^\infty(\mathbb{R}^m, \overline{\mathcal{M}}_{k-1}u), \quad T_{k,r}^* = D_x(I - P_{k,r});$$

$$T_{k,r} : C^\infty(\mathbb{R}^m, \overline{\mathcal{M}}_{k-1}u) \longrightarrow C^\infty(\mathbb{R}^m, \overline{\mathcal{M}}_k), \quad T_k = D_x P_{k,r};$$

$$Q_{k,r} : C^\infty(\mathbb{R}^m, \overline{\mathcal{M}}_{k-1}u) \longrightarrow C^\infty(\mathbb{R}^m, \overline{\mathcal{M}}_{k-1}u), \quad Q_k = D_x(I - P_{k,r}).$$

4.1 Conformal Invariance

We cannot prove conformal invariance and intertwining operators of Q_k with the assumption that D_x is conformally invariant. Here, we correct this using similar techniques that we used in Sect. 3 for the Rarita-Schwinger operators.

Following our Iwasawa decomposition we only need to show the conformal invariance of Q_k under inversion. We also need Cauchy's theorem for the Q_k operator.

Theorem 6 ([14], Stokes' Theorem for Q_k operator) *Let Ω' and Ω be domains in $\mathbb{R}^m$ and suppose the closure of Ω lies in Ω'. Further suppose the closure of Ω is compact and the boundary of Ω, $\partial\Omega$ is piecewise smooth. Then for f, $g \in C^1(\Omega', \mathcal{M}_{k-1})$, we have*

$$\int_\Omega [(g(x, u)uQ_{k,r}, uf(x, u))_u + (g(x, u)u, Q_k uf(x, u))_u]dx^m$$

$$= \int_{\partial\Omega} (g(x, u)u, (I - P_k)d\sigma_x uf(x, u))_u$$

$$= \int_{\partial\Omega} (g(x, u)ud\sigma_x(I - P_{k,r}), uf(x, u))_u$$

where P_k and $P_{k,r}$ are the left and right projections, $d\sigma_x = n(x)d\sigma(x)$, $d\sigma(x)$ is the area element. $(P(u), Q(u))_u = \int_{\mathbb{S}^{m-1}} P(u)Q(u)dS(u)$ is the inner product for any pair of $\mathcal{C}l_m$-valued polynomials.

When $g(x, u)uQ_{k,r} = Q_k uf(x, u) = 0$, we get Cauchy's theorem for Q_k.

Corollary 2 ([14], Cauchy's Theorem for Q_k Operator) *If $Q_k uf(x, u) = 0$ and* $ug(x, u)Q_{k,r} = 0$ *for* $f, g \in C^1(, \Omega', \mathcal{M}_{k-1})$, *then*

$$\int_{\partial\Omega} (g(x, u)u, (I - P_k)d\sigma_x uf(x, u))_u = 0$$

The conformal invariance of the equation $Q_k uf = 0$ under inversion is as follows

Theorem 7 *For any fixed $x \in U \subset \mathbb{R}^m$, let $f(x, u)$ be a left monogenic polynomial homogeneous of degree $k - 1$ in u. If $Q_{k,u}uf(x, u) = 0$, then $Q_{k,w}G(y)\dfrac{ywy}{\|y\|^2}$* $f(y^{-1}, \dfrac{ywy}{\|y\|^2}) = 0$, *where $G(y) = \dfrac{y}{\|y\|^m}$, $x = y^{-1}$, $u = \dfrac{ywy}{\|y\|^2} \in \mathbb{R}^m$.*

Proof First, in Cauchy's theorem, we let $ug(x, u)Q_{k,r} = Q_k uf(x, u) = 0$. Then we have

$$0 = \int_{\partial\Omega} \int_{\mathbb{S}^{m-1}} g(u)u(I - P_k)n(x)uf(x, u)dS(u)d\sigma(x)$$

Let $x = y^{-1}$, we have

$$= \int_{\partial\Omega^{-1}} \int_{\mathbb{S}^{m-1}} g(u)u(I - P_{k,u})G(y)n(y)G(y)uf(y^{-1}, u)dS(u)d\sigma(y),$$

where $G(y) = \dfrac{y}{\|y\|^m}$. Set $u = \dfrac{ywy}{\|y\|^2}$, since $I - P_{k,u}$ interchanges with $G(y)$ [7], we have

$$= \int_{\partial\Omega^{-1}} \int_{\mathbb{S}^{m-1}} g(\frac{ywy}{\|y\|^2})\frac{ywy}{\|y\|^2}G(y)(I - P_{k,w})n(y)G(y)\frac{ywy}{\|y\|^2}f(y^{-1}, \frac{ywy}{\|y\|^2})dS(w)d\sigma(y)$$

$$= \int_{\partial\Omega^{-1}} (g(\frac{ywy}{\|y\|^2})\frac{ywy}{\|y\|^2}G(y), (I - P_{k,w})d\sigma_y G(y)\frac{ywy}{\|y\|^2}f(y^{-1}, \frac{ywy}{\|y\|^2}))_w.$$

According to Stokes' theorem for Q_k,

$$= \int_{\Omega^{-1}} (g(\frac{ywy}{\|y\|^2})\frac{ywy}{\|y\|^2}G(y), Q_{k,w}G(y)\frac{ywy}{\|y\|^2}f(y^{-1}, \frac{ywy}{\|y\|^2}))_w$$

$$+ \int_{\Omega^{-1}} (g(\frac{ywy}{\|y\|^2})\frac{ywy}{\|y\|^2}G(y)Q_{k,w}, G(y)\frac{ywy}{\|y\|^2}f(y^{-1}, \frac{ywy}{\|y\|^2}))_w.$$

Since $ug(x, u)$ is arbitrary in the kernel of $Q_{k,r}$ and $uf(x, u)$ is arbitrary in the kernel of Q_k, we get $g(\frac{ywy}{\|y\|^2})\frac{ywy}{\|y\|^2}G(y)Q_{k,w} = Q_{k,w}G(y)\frac{ywy}{\|y\|^2}f(y^{-1}, \frac{ywy}{\|y\|^2}) = 0$.

To complete this section, we provide *Stokes' theorem* for other Rarita-Schwinger type operators as follows:

Theorem 8 (Stokes' Theorem for T_k) *Let Ω' and Ω be domains in $\mathbb{R}^m$ and suppose the closure of Ω lies in Ω'. Further suppose the closure of Ω is compact and $\partial\Omega$ is piecewise smooth. Let $f, g \in C^1(\Omega', \mathcal{M}_k)$. Then*

$$\int_{\Omega} \left[(g(x, u)T_k, f(x, u))_u + (g(x, u), T_k f(x, u)) \right] dx^m$$

$$= \int_{\partial\Omega} (g(x, u), P_k d\sigma_x f(x, u))_u$$

$$= \int_{\partial\Omega} (g(x, u)d\sigma_x P_{k,r}, f(x, u))_u,$$

where P_k and $P_{k,r}$ are the left and right projections, $d\sigma_x = n(x)d\sigma(x)$ and $(P(u), Q(u))_u = \int_{\mathbb{S}^{m-1}} P(u)Q(u)dS(u)$ is the inner product for any pair of $\mathcal{C}l_m$-valued polynomials.

Theorem 9 (Stokes' Theorem for T_k^*) *Let Ω' and Ω be domains in $\mathbb{R}^m$ and suppose the closure of Ω lies in Ω'. Further suppose the closure of Ω is compact and $\partial\Omega$ is piecewise smooth. Let $f, g \in C^1(\Omega', u\mathcal{M}_{k-1})$. Then*

$$\int_{\Omega} \left[(g(x, u)T_k^*, f(x, u))_u + (g(x, u), T_k^* f(x, u)) \right] dx^m$$

$$= \int_{\partial\Omega} (g(x, u), (I - P_k)d\sigma_x f(x, u))_u$$

$$= \int_{\partial\Omega} (g(x, u)d\sigma_x(I - P_{k,r}), f(x, u))_u,$$

where P_k and $P_{k,r}$ are the left and right projections, $d\sigma_x = n(x)d\sigma(x)$ and $(P(u), Q(u))_u = \int_{\mathbb{S}^{m-1}} P(u)Q(u)dS(u)$ is the inner product for any pair of $\mathcal{C}l_m$-valued polynomials.

Theorem 10 (Alternative Form of Stokes' Theorem) *Let Ω and Ω' be as in the previous theorem. Then for $f \in C^1(\mathbb{R}^m, \mathcal{M}_k)$ and $g \in C^1(\mathbb{R}^m, \mathcal{M}_{k-1})$, we have*

$$\int_{\partial\Omega} \left(g(x, u)ud\sigma_x f(x, u) \right)_u$$

$$= \int_{\Omega} \left(g(x, u)uT_k, f(x, u) \right)_u dx^m + \int_{\Omega} \left(g(x, u)u, T_k^* f(x, u) \right)_u dx^m.$$

Further

$$\int_{\partial\Omega} \left(g(x, u)u d\sigma_x f(x, u)\right)_u$$

$$= \int_{\partial\Omega} \left(g(x, u)u, (I - P_k)d\sigma_x f(x, u)\right)_u$$

$$= \int_{\partial\Omega} \left(g(x, u)u d\sigma_x P_k, f(x, u)\right)_u.$$

Acknowledgements The authors wish to thank the referee for helpful suggestions that improved the manuscript. The authors are also grateful to Bent Ørsted for communications pointing out that the intertwining operators for the Rarita-Schwinger operators are special cases of Knapp-Stein intertwining operators in higher spin theory [5, 12].

References

1. Atiyah, M.F., Bott, R., Shapiro, A.: Clifford modules. Topology 3(Suppl. 1), 3–38 (1964)
2. De Bie, H., Eelbode, D., Roels, M.: The higher spin Laplace operator. Potential Anal. **47**(2), 123–149 (2017)
3. Brackx, F., Delanghe, R., Sommen, F.: Clifford Analysis. Pitman, London (1982)
4. Bureš, J., Sommen, F., Souček, V., Van Lancker, P.: Rarita-schwinger type operators in clifford analysis. J. Funct. Anal. **185**(2), 425–455 (2001)
5. Clerc, J.L., Orsted, B.: Conformal covariance for the powers of the Dirac operator. arXiv:1409.4983
6. Delanghe, R., Sommen, F., Souček, V.: Clifford Algebra and Spinor-Valued Functions: A Function Theory for the Dirac Operator. Kluwer, Dordrecht (1992)
7. Dunkl, C.F., Li, J., Ryan, J., Van Lancker, P.: Some Rarita-Schwinger type operators. Comput. Methods Funct. Theor. **13**(3), 397–424 (2013)
8. Eastwood, M.: The Cartan product. Bull. Belgian Math. Soc. **11**(5), 641–651 (2005)
9. Eastwood, M.G., Ryan, J.: Aspects of dirac operators in analysis. Milan J. Math. **75**(1), 91–116 (2007)
10. Gilbert, J., Murray, M.: Clifford Algebras and Dirac Operators in Harmonic Analysis. Cambridge University Press, Cambridge (1991)
11. Humphreys, J.E.: Introduction to Lie algebras and Representation Theory, Graduate Texts in Mathematics, Readings in Mathematics 9. Springer, New York (1972)
12. Knapp, A.W., Stein, E.M.: Intertwining operators for semisimple groups. Ann. Math. **93**(3), 489–578 (1971)
13. Van Lancker, P., Sommen, F., Constales, D.: Models for irreducible representations of Spin(m). Ad. Appl. Clifford Algebras **11**(1 supplement), 271–289 (2001)
14. Li, J., Ryan, J.: Some operators associated to Rarita-Schwinger type operators. Complex Var. Elliptic Equ. Int. J. **57**(7–8), 885–902 (2012)
15. Lounesto, P.: Clifford Algebras and Spinors, London Mathematical Society Lecture Note Series 286, Cambridge University Press (2001)
16. Porteous, I.: Clifford Algebra and the Classical Groups. Cambridge University Press, Cambridge (1995)
17. Roels, M.: A Clifford analysis approach to higher spin fields. Master Thesis, University of Antwerp (2013)
18. Ryan, J.: Conformally coinvariant operators in Clifford analysis. Z. Anal. Anwendungen **14**, 677–704 (1995)

19. Ryan, J.: Iterated Dirac operators and conformal transformations in $\mathbb{R}^m$. In: Proceedings of the XV International Conference on Differential Geometric Methods in Theoretical Physics, World Scientific, pp. 390–399 (1987)
20. Shirrell, S.: Hermitian Clifford Analysis and its connections with representation theory. Bachelor Thesis (2011)
21. Stein, E., Weiss, G.: Generalization of the Cauchy-Riemann equations and representations of the rotation group. Amer. J. Math. **90**, 163–196 (1968)

Notions of Regularity for Functions of a Split-Quaternionic Variable

J. A. Emanuello and C. A. Nolder

Abstract The utility and beauty of the theory holomorphic functions of a complex variable leads one to wonder whether analogous function theories exist for other (presumably higher dimensional) algebras. Over the last several decades it has been shown that much of Complex analysis extends to a similar theory for the family of Clifford Algebras $C\ell_{0,n}$. However, there has yet to be a complete description for the general theory over the family of Clifford algebras $C\ell_{p,q}$ (for $p \neq 0$). In this work, we describe two different approaches from the literature for finding a theory of "holomorphic" functions of a split-quaternionic variable (which is the Clifford algebra $C\ell_{1,1}$). We show that one approach yields a relatively trivial theory, while the other gives a richer one. In the second instance, we describe a simple subclass of "holomorphic" functions and give two examples of an analogue of the Cauchy-Kowalewski extension in this context.

Keywords Split-quaternionic variables · Clifford analysis

1 Introduction

One need look no further than a text on complex analysis, such as [1] or especially [2], to know that algebraic properties of $\mathbb{C}$ play a major role in the analysis and geometry of the plane. The simple fact that $i^2 = -1$ gives rise to the Cauchy-Riemann equations, which is the foundation of the theory of holomorphic functions, which are those

J. A. Emanuello · C. A. Nolder (✉)
Department of Mathematics, The Florida State University, 208 Love Building,
1017 Academic Way, Tallahassee, FL 32306-4510, USA
e-mail: nolder@math.fsu.edu; craiganolder@hotmail.com

J. A. Emanuello
e-mail: jemanuel@math.fsu.edu

© Springer Nature Switzerland AG 2018
P. Cerejeiras et al. (eds.), *Clifford Analysis and Related Topics*,
Springer Proceedings in Mathematics & Statistics 260,
https://doi.org/10.1007/978-3-030-00049-3_5

functions of a complex variable which are differentiable in a complex sense. Indeed, the existence of the limit of the difference quotient

$$\lim_{\Delta z \to 0} \frac{f(z + \Delta z) - f(z)}{\Delta z}$$

means that the limit is the same whether $\Delta z = \Delta x$ or $\Delta z = i \Delta y$. That is,

$$\frac{\partial u}{\partial x} + i \frac{\partial v}{\partial x} = \frac{1}{i} \frac{\partial u}{\partial y} + \frac{\partial v}{\partial y},$$

and the C-R equations are obtained:

$$\frac{\partial u}{\partial x} = \frac{\partial v}{\partial y} \text{ and } \frac{\partial v}{\partial x} = -\frac{\partial u}{\partial y}.$$

The minus sign in the second equation occurs because $\frac{1}{i} = -i$, which is a direct consequence of $i^2 = -1$. Thus, when a function of a complex variable with C^1 components is holomorphic if and only if the C-R equations are satisfied.

One may also consider functions of a complex variable which are annihilated by the operator

$$\partial_{\bar{z}} := \frac{1}{2} \left(\frac{\partial}{\partial x} + i \frac{\partial}{\partial y} \right).$$

Indeed, a C^1 function is holomorphic if and only if it is annihilated by $\partial_{\bar{z}}$ and its complex derivative is given by $\partial_z f$, where

$$\partial_z := \frac{1}{2} \left(\frac{\partial}{\partial x} - i \frac{\partial}{\partial y} \right).$$

For decades, many mathematicians have been interested in extending complex analysis to new settings involving higher dimensional analogues of $\mathbb{C}$. One such way is to consider functions valued in a Clifford algebra. Informally speaking, Clifford algebras are associative algebras with unit and an embedded vector space (normally, this is a euclidean or pseudo-euclidean space [3]). Examples include the complex numbers and the quaternions.

Although usually credited to Sir William Rowan Hamilton, the famous Irish mathematician, the quaternions were actually discovered in 1840 by Benjamin Olinde Rodrigues, a Frenchman who never received credit at the time (and rarely receives credit today) [4]. Three years later, Hamilton would discover independently the algebra after years of trying to find higher dimensional analogues of $\mathbb{C}$ [5].

Thirty-five years later, William Kingdon Clifford, an English Geometer and Philosopher, combined the work of Hamilton (there is no evidence he was aware of Rodrigues's work) and Grassmann to create what he called *geometric algebras* [6]. They were since renamed *Clifford algebras* and have been used by physicists and mathematicians alike. Of notable importance is P. Dirac's use of the γ- matrices (see the original work in [7]), which are the generators of the Clifford algebra $C\ell_{1,3}$, to linearize the Klein-Gordan equation [8]. The literature is full of interesting facts about Clifford algebras, and the reader is encouraged to read [3, 9] for additional background.

Indeed, much of complex analysis transfers to a Clifford algebra setting, especially for the so-called euclidean Clifford algebras $C\ell_{0,n}$. In fact, functions valued in $C\ell_{0,n}$ which are annihilated by an analogue of the Cauchy-Riemann operator satisfy many desired properties (see [9, 10] for details). Additionally, the quaternions are considered as a special case in [11].

Unlike in the complex case, when we consider functions of a split-quaternionic variable (i.e. those functions with domain and range contained in the Clifford algebra $C\ell_{1,1}$) and explore the two analogous ways of defining a holomorphic function, we find that they are not equivalent. Thus, two different theories of holomorphic functions can be studied, as in [12, 13]. However, the one in [12] stands out as the more natural analogue because it gives rise to a (relatively) large class of functions to be studied. Indeed, for the analogue defined in [13] we show (by adopting a proof of an analogous statement in Sudbery's paper [11]) that only affine functions, which is a (relatively) small class of functions, satisfy the given conditions.

1.1 The Split-Quaternions

The split-quaternions are the real Clifford algebra

$$C\ell_{1,1} := \{Z = x_0 + x_1 i + x_2 j + x_3 ij \ : \ x_0, x_1, x_2, x_3 \in \mathbb{R}\}.$$

Though they resemble Hamilton's quaternions, the multiplication rules in $C\ell_{1,1}$ are as follows:

$$ij = -ji \text{ and } \varepsilon^2 = \begin{cases} -1 & \text{if } \varepsilon = i \\ 1 & \text{if } \varepsilon = j. \end{cases}$$

Thus, if $A = a + ib + jc + ijd$, $Z = x_0 + x_1 i + x_2 j + x_3 ij \in C\ell_{1,1}$, then

$$AZ = (ax_0 - bx_1 + cx_2 + dx_3) + i(bx_0 + ax_1 + dx_2 - cx_3)$$
$$+ j(cx_0 + dx_1 + ax_2 - bx_3) + ij(dx_0 - cx_1 + bx_2 + ax_3).$$

One may define an involution analogous to complex conjugation as follows:

$$\overline{Z} = x_0 - x_1 i - x_2 j - x_3 ij$$

In a manner similar to the split-complex case [14, 15], we may obtain the indefinite quadratic form $Q_{2,2}$ by

$$\langle Z \rangle := Z\overline{Z} = x_0^2 + x_1^2 - x_2^2 - x_3^2.$$

Hence we shall identify the split-quaternions with $\mathbb{R}^{2,2}$. It is easily shown that when Z^{-1} exists, it is given by

$$Z^{-1} = \frac{\overline{Z}}{\langle Z \rangle}.$$

Indeed, Z^{-1} exists if and only if $\langle Z \rangle \neq 0$ (and hence is a zero divisor otherwise). Indeed, Z is a zero divisor if and only if

$$x_0^2 + x_1^2 = x_2^2 + x_3^2.$$

One way this condition is satisfied is if Z is of the form

$$Z = r \cos\theta + r \sin\theta i + r \cos\phi j + r \sin\phi ij,$$

where $r \in \mathbb{R}$ and $\theta, \phi \in [0, 2\pi)$. Below, we shall see that there are other classes of zero divisors.

To that end, we shall take a minor, but fruitful detour to idempotents, of which there are uncountably many.

Proposition 1 *Suppose* $Z = ri + (\sqrt{1+r^2}\cos\theta)j + (\sqrt{1+r^2}\sin\theta)ij$. *Then* $\frac{1}{2}(1 \pm Z)$ *is an idempotent.*

Proof The proof is a corollary of a similar fact found [9], which essentially says that for every element of order 2 we get two idempotents. It can be show that Z is order 2 if and only if

$$\begin{cases} x_0^2 - x_1^2 + x_2^2 + x_3^2 &= 1 \\ x_0 x_1 &= 0 \\ x_0 x_2 &= 0 \\ x_0 x_3 &= 0 \end{cases}.$$

Assuming Z is not real, then it follows that

$$x_2^2 + x_3^2 = 1 + x_1^2.$$

Setting $x_1 = r \in \mathbb{R}$, we must have that

$$Z = ri + (\sqrt{1+r^2}\cos\theta)j + (\sqrt{1+r^2}\sin\theta)ij,$$

where and $\theta \in [0, 2\pi)$.

Now, if $Z^2 = 1$, then

$$\left(\frac{1}{2}(1 \pm Z)\right)^2 = \frac{1}{4}(1 \pm 2Z + Z^2)$$
$$= \frac{1}{4}(2 \pm 2Z)$$
$$= \frac{1}{2}(1 \pm Z).$$

Remark 1 This is in contrast with the complex and split-complex cases, which have two (0 and 1) and four idempotents (0, 1, and $1 \pm j$) respectively. In fact, such an idempotent is a zero divisor, for if $Z^2 = 1$ then

$$\frac{(1+Z)}{2} \cdot \frac{(1-Z)}{2} = \frac{(1-Z^2)}{4} = 0.$$

Furthermore, such a zero divisor is of a different form than the ones discussed above.

Lastly, we discuss some other representations of the split-quaternions. Indeed, there are a number of ways to express the split-quaternions as 2×2 matrices over $\mathbb{R}$ and $\mathbb{C}$.

Lemma 1 *As algebras, the split-quaternions and real 2×2 matrices are isomorphic.*

Proof If we identify

$$1 \sim \begin{bmatrix} 1 & 0 \\ 0 & 1 \end{bmatrix}, \ i \sim \begin{bmatrix} 0 & -1 \\ 1 & 0 \end{bmatrix}, \ j \sim \begin{bmatrix} 0 & 1 \\ 1 & 0 \end{bmatrix},$$

then we may map $C\ell_{1,1}$ to the real 2×2 matrices by

$$x_0 + x_1 i + x_2 j + x_3 ij \longmapsto \begin{bmatrix} x_0 + x_3 & -x_1 + x_2 \\ x_1 + x_2 & x_0 - x_3 \end{bmatrix}.$$

Notice that

$$\det \begin{bmatrix} x_0 + x_3 & -x_1 + x_2 \\ x_1 + x_2 & x_0 - x_3 \end{bmatrix} = x_0^2 + x_1^2 - x_2^2 - x_3^2,$$

which is the form $Q_{2,2}$. It's easy to check that this gives an algebra homomorphism. Further,

$$\begin{bmatrix} y_1 & y_2 \\ y_3 & y_4 \end{bmatrix} \longmapsto \frac{1}{2}[(y_1 + y_4) + (y_3 - y_2)i + (y_3 + y_2)j + (y_1 - y_4)ij]$$

gives a two-sided inverse, so that the above is an algebra isomorphism.

Functions of a split-quaternionic variable and notions of regularity have been the subject of interest in the literature [12, 13]. It is worth noting that the split-quaternions contain both the complex numbers, realized as the Clifford algebra $C\ell_{0,1} := \{x_0 + x_1 i \ : \ x_0, x_1 \in \mathbb{R}\}$ and split-complex numbers, $C\ell_{1,0} := \{x_0 + x_2 j \ : \ x_0, x_2 \in \mathbb{R}\}$, as subalgebras. For more information about the latter, please see [14–20].

2 Notions of Holomorphy

The functions we are concerned with are

$$f : U \subseteq \mathbb{R}^{2,2} \to C\ell_{1,1},$$

where U is open (in the euclidean sense). As higher dimensional analogues of functions of a complex variable, we are interested in obtaining an analogous definition for *holomorphic functions*. As we shall see, there are various ways of doing this in the literature.

The first and most interesting way is through split quaternionic valued differential operators [12]. The second is more recent and less interesting and is obtained by considering a difference quotient [13].

2.1 Analogues of the Cauchy-Riemann Operator

Recall that in complex analysis, one considers the Dirac operators $\partial_{\bar{z}}$ and ∂_z, whose product (in either order) gives the Laplacian for $\mathbb{R}^2$, usually denoted by Δ. Of course, f is called holomorphic if $\partial_{\bar{z}} f = 0$ and its (complex) derivative is given by $\partial_z f$. Additionally, the real and imaginary parts of f are harmonic functions, and the Dirichlet problem is well-posed.

The question asked in the literature is: *Can we define operators valued in $C\ell_{1,1}$ which resemble $\partial_{\bar{z}}$ and ∂_z?* This question has been answered in the affirmative, although with little mention of the differential geometry which lies just below the surface.

However the question we are really asking is: can we factorize the Laplacian in $\mathbb{R}^{2,2}$ with linear first order operators over $C\ell_{1,1}$? In this semi-Riemannian manifold, the Laplacian, which is understood to be the derivative of the gradient, is given by [21]:

$$\Delta_{2,2} = \frac{\partial^2}{\partial x_0^2} + \frac{\partial^2}{\partial x_1^2} - \frac{\partial^2}{\partial x_2^2} - \frac{\partial^2}{\partial x_3^2}.$$

It is easy to check that the linear operators

$$\overline{\partial} := \frac{\partial}{\partial x_0} + i\frac{\partial}{\partial x_1} - j\frac{\partial}{\partial x_2} - ij\frac{\partial}{\partial x_3} \text{ and}$$

$$\partial := \frac{\partial}{\partial x_0} - i\frac{\partial}{\partial x_1} + j\frac{\partial}{\partial x_2} + ij\frac{\partial}{\partial x_3}$$

are factors of $\Delta_{2,2}$. Due to the non-commutativity of $C\ell_{1,1}$, these operators may be applied to functions on either the left or right and with different results, in general.

Remark 2 There are other factorizations of $\Delta_{2,2}$ inside $C\ell_{1,1}$. Our choice of $\overline{\partial}$ is deliberate– it is the gradient inside the semi-Riemannian manifold $\mathbb{R}^{2,2}$ (just as the Cauchy-Riemann operator $\partial_{\overline{z}}$ resembles the gradient in $\mathbb{R}^2$). For alternative interpretation of this idea, see [22].

In [23], it is carefully explained that the factorization of $\Delta_{2,2}$ given above is compatible with a preferred factorization of the Laplace operator in four dimensions over the quaternions.

Definition 1 Let $U \subset C\ell_{1,1} \cong \mathbb{R}^{2,2}$ and let $F : U \to C\ell_{1,1}$ be $C^1(U)$. We say F is left regular if

$$\overline{\partial} F = 0$$

for every $Z \in U$. Similarly, we say F is right regular if

$$F\overline{\partial} = 0$$

for every $Z \in U$.

We have adopted the above definition from [12], which contains a proof of a Cauchy-like integral formula for left-regular functions.

By multiplying arbitrary F with $\overline{\partial}$ we obtain the following conditions which make it easier to check left and right regularity.

Proposition 2 *Let $F : U \to C\ell_{1,1}$ be $C^1(U)$. Then F is left regular if and only if it satisfies the system of PDEs:*

$$\begin{cases} \dfrac{\partial f_0}{\partial x_0} - \dfrac{\partial f_1}{\partial x_1} - \dfrac{\partial f_2}{\partial x_2} - \dfrac{\partial f_3}{\partial x_3} = 0 \\[2mm] \dfrac{\partial f_1}{\partial x_0} + \dfrac{\partial f_0}{\partial x_1} + \dfrac{\partial f_3}{\partial x_2} - \dfrac{\partial f_2}{\partial x_3} = 0 \\[2mm] \dfrac{\partial f_2}{\partial x_0} - \dfrac{\partial f_3}{\partial x_1} - \dfrac{\partial f_0}{\partial x_2} - \dfrac{\partial f_1}{\partial x_3} = 0 \\[2mm] \dfrac{\partial f_3}{\partial x_0} + \dfrac{\partial f_2}{\partial x_1} + \dfrac{\partial f_1}{\partial x_2} - \dfrac{\partial f_0}{\partial x_3} = 0. \end{cases}$$

Proof The proof follows directly from the definition. Simply multiply in the proper order, collect like components together, and equate them to zero to obtain the desired system.

Proposition 3 *Let* $F : U \to C\ell_{1,1}$ *be* $C^1(U)$. *Then* F *is right regular if and only if it satisfies the system of PDEs:*

$$\begin{cases} \dfrac{\partial f_0}{\partial x_0} - \dfrac{\partial f_1}{\partial x_1} - \dfrac{\partial f_2}{\partial x_2} - \dfrac{\partial f_3}{\partial x_3} = 0 \\[2mm] \dfrac{\partial f_1}{\partial x_0} + \dfrac{\partial f_0}{\partial x_1} - \dfrac{\partial f_3}{\partial x_2} + \dfrac{\partial f_2}{\partial x_3} = 0 \\[2mm] \dfrac{\partial f_2}{\partial x_0} + \dfrac{\partial f_3}{\partial x_1} - \dfrac{\partial f_0}{\partial x_2} + \dfrac{\partial f_1}{\partial x_3} = 0 \\[2mm] \dfrac{\partial f_3}{\partial x_0} - \dfrac{\partial f_2}{\partial x_1} - \dfrac{\partial f_1}{\partial x_2} - \dfrac{\partial f_0}{\partial x_3} = 0. \end{cases}$$

However, this notion of regularity does not produce a completely parallel function theory, in that $C\ell_{1,1}$ analogues of basic holomorphic functions in the complex plane are not regular.

Example 1 Let $A = a + ib + jc + ijd \in C\ell_{1,1}$. Then recall

$$AZ = (ax_0 - bx_1 + cx_2 + dx_3) + i(bx_0 + ax_1 + dx_2 - cx_3)$$
$$+ j(cx_0 + dx_1 + ax_2 - bx_3) + ij(dx_0 - cx_1 + bx_2 + ax_3).$$

Thus,

$$\bar{\partial}(AZ) = (a + ib + jc + dij) + i(-b + ai + dj - cij)$$
$$- j(c + id + aj + bij) - ij(d - ic - bj + aij)$$
$$= -2a + i2b + j2c + ij2d$$
$$= -2\bar{A},$$

A similar calculation shows that

$$(AZ)\bar{\partial} = -2A.$$

Other calculations show that the function ZA is also neither left-regular nor right-regular.

It is also the case that the function

$$F(Z) = Z^2 = (x_0^2 - x_1^2 + x_2^2 + x_3^2) + ix_0x_1 + jx_0x_2 + ijx_0x_3$$

is neither left or right regular.

Remark 3 It should be noted that the above issue is not unique to the split-quaternions. Similar functions over the quaternions (or even higher dimensional Clifford algebras) have similar issues.

We obtain similar systems of PDEs if we consider the equations $\partial F = 0$ and $F\partial = 0$:

$$\begin{cases} \dfrac{\partial f_0}{\partial x_0} + \dfrac{\partial f_1}{\partial x_1} + \dfrac{\partial f_2}{\partial x_2} + \dfrac{\partial f_3}{\partial x_3} = 0 \\[3mm] \dfrac{\partial f_1}{\partial x_0} - \dfrac{\partial f_0}{\partial x_1} - \dfrac{\partial f_3}{\partial x_2} + \dfrac{\partial f_2}{\partial x_3} = 0 \\[3mm] \dfrac{\partial f_2}{\partial x_0} + \dfrac{\partial f_3}{\partial x_1} - \dfrac{\partial f_0}{\partial x_2} + \dfrac{\partial f_1}{\partial x_3} = 0 \\[3mm] \dfrac{\partial f_3}{\partial x_0} - \dfrac{\partial f_2}{\partial x_1} - \dfrac{\partial f_1}{\partial x_2} + \dfrac{\partial f_0}{\partial x_3} = 0 \end{cases}$$

and

$$\begin{cases} \dfrac{\partial f_0}{\partial x_0} + \dfrac{\partial f_1}{\partial x_1} + \dfrac{\partial f_2}{\partial x_2} + \dfrac{\partial f_3}{\partial x_3} = 0 \\[3mm] \dfrac{\partial f_1}{\partial x_0} - \dfrac{\partial f_0}{\partial x_1} + \dfrac{\partial f_3}{\partial x_2} - \dfrac{\partial f_2}{\partial x_3} = 0 \\[3mm] \dfrac{\partial f_2}{\partial x_0} - \dfrac{\partial f_3}{\partial x_1} + \dfrac{\partial f_0}{\partial x_2} - \dfrac{\partial f_1}{\partial x_3} = 0 \\[3mm] \dfrac{\partial f_3}{\partial x_0} + \dfrac{\partial f_2}{\partial x_1} + \dfrac{\partial f_1}{\partial x_2} + \dfrac{\partial f_0}{\partial x_3} = 0. \end{cases}$$

These also fail to produce analogues of holomorphy whereby affine functions and the squaring function fail these conditions.

Example 2 Let $A = a + ib + jc + ijd \in C\ell_{1,1}$. Then,

$$\begin{aligned} \partial(AZ) &= (a + ib + jc + dij) - i(-b + ai + dj - cij) \\ &\quad + j(c + id + aj + bij) + ij(d - ic - bj + aij) \\ &= 4a. \end{aligned}$$

A similar calculation shows that

$$(AZ)\,\partial = 4A.$$

Other calculations show that the function ZA is not annihilated by ∂ on either side. The same is true for the function $F(Z) = Z^2$.

Remark 4 Notice that if $\overline{\partial} F = 0$, then $\overline{\overline{\partial} F} = 0$. But by properties of the conjugate, $\overline{\partial F} = \overline{F}\partial$. That is,

$$\overline{\partial} F = 0 \iff \overline{F}\partial = 0.$$

Similarly,

$$\partial F = 0 \iff \overline{F}\left(\overline{\partial}\right) = 0.$$

Thus, there is no need to consider a separate theory by multiplying operators on the right.

2.2   Difference Quotients

Recall, another (and probably primary) way to define holomorphic functions of a complex variable is via the limit of a difference quotient:

$$\lim_{\Delta z \to 0} \frac{f(z + \Delta z) - f(z)}{\Delta z}.$$

One obtains the Cauchy-Riemann equations by allowing Δz to approach 0 along the real axis and again along the imaginary axis and then setting the results equal to each other.

In Masouri et. al., a similar method is used to produce another analogue of holomorphy [13]. However, since the split-quaternions are not commutative, so there are two ways to construct an analogue of the difference quotient. In Masouri the "quotient" is defined by

$$\lim_{\substack{\Delta Z \to 0 \\ \Delta Z \text{ invertible}}} (f(Z + \Delta Z) - f(Z))(\Delta Z)^{-1}.$$

When this limit exists, such functions are called *right $C\ell_{1,1}$-differentiable*, or just *right-differentiable*. By setting ΔZ equal to Δx_0, $i\,\Delta x_1$, $j\,\Delta x_2$, and $ij\,\Delta x_3$, taking the limit in each instance, we get four ways to take the "derivative" [13]. That is,

$$\lim_{\Delta x_0 \to 0} (f(Z + \Delta x_0) - f(Z))(\Delta x_0)^{-1} = \frac{\partial f_0}{\partial x_0} + i\frac{\partial f_1}{\partial x_0} + j\frac{\partial f_2}{\partial x_0} + ij\frac{\partial f_3}{\partial x_0},$$

$$\lim_{i\Delta x_1 \to 0} (f(Z + i\Delta x_1) - f(Z))(i\Delta x_1)^{-1} = -i\frac{\partial f_0}{\partial x_1} + \frac{\partial f_1}{\partial x_1} + ij\frac{\partial f_2}{\partial x_1} - j\frac{\partial f_3}{\partial x_1},$$

$$\lim_{j\Delta x_2 \to 0} (f(Z + j\Delta x_2) - f(Z))(j\Delta x_2)^{-1} = j\frac{\partial f_0}{\partial x_2} + ij\frac{\partial f_1}{\partial x_2} + \frac{\partial f_2}{\partial x_2} + i\frac{\partial f_3}{\partial x_2},$$

and

$$\lim_{ij\Delta x_3 \to 0} (f(Z + ij\Delta x_3) - f(Z))(ij\Delta x_3)^{-1} = ij\frac{\partial f_0}{\partial x_3} - j\frac{\partial f_1}{\partial x_3} - i\frac{\partial f_2}{\partial x_3} + \frac{\partial f_3}{\partial x_3}.$$

Equating the four results, we obtain the system of PDEs [13]:

$$\begin{cases} \dfrac{\partial f_0}{\partial x_0} = \dfrac{\partial f_1}{\partial x_1} = \dfrac{\partial f_2}{\partial x_2} = \dfrac{\partial f_3}{\partial x_3} \\[2mm] \dfrac{\partial f_1}{\partial x_0} = -\dfrac{\partial f_0}{\partial x_1} = \dfrac{\partial f_3}{\partial x_2} = -\dfrac{\partial f_2}{\partial x_3} \\[2mm] \dfrac{\partial f_2}{\partial x_0} = -\dfrac{\partial f_3}{\partial x_1} = \dfrac{\partial f_0}{\partial x_2} = -\dfrac{\partial f_1}{\partial x_3} \\[2mm] \dfrac{\partial f_3}{\partial x_0} = \dfrac{\partial f_2}{\partial x_1} = \dfrac{\partial f_1}{\partial x_2} = \dfrac{\partial f_0}{\partial x_3}. \end{cases}$$

Although the work which introduces this notion of differentiability [13], does not mention any specific examples of right $C\ell_{1,1}$-differentiable functions, the entire class of these functions are merely the affine mappings.

Example 3 Recall that

$$AZ + K = (ax_0 - bx_1 + cx_2 + dx_3 + k) + i(bx_0 + ax_1 + dx_2 - cx_3 + \ell)$$
$$+ j(cx_0 + dx_1 + ax_2 - bx_3 + m) + ij(dx_0 - cx_1 + bx_2 + ax_3 + n).$$

Notice $f(Z) = AZ + K$ clearly satisfies the requisite system of PDEs and is, thus, right $C\ell_{1,1}$-differentiable. Indeed, the "derivative" is

$$\lim_{\substack{\Delta Z \to 0 \\ \Delta Z \text{ invertible}}} (A(Z + \Delta Z) + K - AZ - K)(\Delta Z)^{-1}$$

$$= \lim_{\substack{\Delta Z \to 0 \\ \Delta Z \text{ invertible}}} (A\Delta Z)(\Delta Z)^{-1} = A.$$

Theorem 1 *Let $U \subseteq \mathbb{R}^{2,2}$ be open and connected and $F : U \to C\ell_{1,1}$. Then F is right $C\ell_{1,1}$-differentiable if and only $F(Z) = AZ + K$, where $A, K \in C\ell_{1,1}$. That is, the right $C\ell_{1,1}$-differentiable functions must be affine mappings.*

Proof As similar fact is true for functions of a quaternionic variable and so we follow a similar proof from Sudbery's paper[1] [11].

First notice that $Z = (x_0 + x_1 i) + (x_2 + x_3 i)j = z + wj$. As such we may write $f(Z) = g(z, w) + h(z, w)j$, where $g(z, w) = f_0(z, w) + if_1(z, w)$ and $h(z, w) = f_2(z, w) + if_3(z, w)$.

Now, the above system of PDEs gives us that g is holomorphic with respect to the complex variables z and $\overline{w}$. Similarly, h is holomorphic with respect to the complex variables w and $\overline{z}$. Additionally,

$$\frac{\partial g}{\partial z} = \frac{\partial f_0}{\partial x_0} + i\frac{\partial f_1}{\partial x_0} = \frac{\partial f_2}{\partial x_2} + i\frac{\partial f_3}{\partial x_2} = \frac{\partial h}{\partial w},$$

$$\frac{\partial g}{\partial \overline{w}} = -\frac{\partial f_1}{\partial x_3} + i\frac{\partial f_0}{\partial x_3} = -\frac{\partial f_3}{\partial x_1} + i\frac{\partial f_2}{\partial x_1} = \frac{\partial h}{\partial \overline{z}}.$$

Now, g and h have continuous partial derivatives of all orders. Thus, we must have

$$\frac{\partial^2 g}{\partial z^2} = \frac{\partial}{\partial z}\left(\frac{\partial h}{\partial w}\right) = \frac{\partial}{\partial w}\left(\frac{\partial h}{\partial z}\right) = 0,$$

$$\frac{\partial^2 h}{\partial w^2} = \frac{\partial}{\partial w}\left(\frac{\partial g}{\partial z}\right) = \frac{\partial}{\partial z}\left(\frac{\partial g}{\partial w}\right) = 0,$$

$$\frac{\partial^2 g}{\partial \overline{w}^2} = \frac{\partial}{\partial \overline{w}}\left(\frac{\partial h}{\partial \overline{z}}\right) = \frac{\partial}{\partial \overline{z}}\left(\frac{\partial h}{\partial \overline{w}}\right) = 0,$$

$$\frac{\partial^2 h}{\partial \overline{z}^2} = \frac{\partial}{\partial \overline{z}}\left(\frac{\partial g}{\partial \overline{w}}\right) = \frac{\partial}{\partial \overline{w}}\left(\frac{\partial g}{\partial \overline{z}}\right) = 0.$$

W.L.O.G. we may assume that U is connected and convex (since each connected component may be covered by convex sets, which overlap pair-wise on convex sets). Thus integrating on line segments allows us to conclude that g and h are linear:

$$g(z, w) = \alpha + \beta z + \gamma \overline{w} + \delta z \overline{w},$$

$$h(z, w) = \epsilon + \eta \overline{z} + \theta w + v \overline{z} w.$$

Since $\frac{\partial g}{\partial z} = \frac{\partial h}{\partial w}$, we must have that $\beta = \theta$ and $\delta = v = 0$. Also since $\frac{\partial g}{\partial \overline{w}} = \frac{\partial h}{\partial \overline{z}}$, it is the case that $\gamma = \eta$. Thus,

---

[1] We are very grateful to Professor Uwe Kähler of University of Aveiro for bringing this paper to our attention.

$$f(Z) = g(z, w) + h(z, w)j$$
$$= (\alpha + \beta z + \gamma \overline{w}) + (\epsilon + \gamma \overline{z} + \beta w)j$$
$$= (\beta + \gamma j)(z + wj) + (\alpha + \epsilon j)$$
$$= AZ + K,$$

as required.

Remark 5 The above theorem proves that right $C\ell_{1,1}$-differentiable functions are not left or right regular and conversely (except for when $A = 0$). Indeed, they are also not annihilated by ∂ on either side.

As an alternative to the definition found in [13], one may reverse the multiplication in the difference quotient to obtain

$$\lim_{\substack{\Delta Z \to 0 \\ \Delta Z \text{ invertible}}} (\Delta Z)^{-1} (f(Z + \Delta Z) - f(Z)).$$

When this limit exists, such functions are called *left $C\ell_{1,1}$-differentiable*. Proceeding as above, a slightly different system of PDEs than the one found in [13] is obtained:

$$\begin{cases} \dfrac{\partial f_0}{\partial x_0} = \dfrac{\partial f_1}{\partial x_1} = \dfrac{\partial f_2}{\partial x_2} = \dfrac{\partial f_3}{\partial x_3} \\[2ex] \dfrac{\partial f_1}{\partial x_0} = -\dfrac{\partial f_0}{\partial x_1} = -\dfrac{\partial f_3}{\partial x_2} = \dfrac{\partial f_2}{\partial x_3} \\[2ex] \dfrac{\partial f_2}{\partial x_0} = \dfrac{\partial f_3}{\partial x_1} = \dfrac{\partial f_0}{\partial x_2} = \dfrac{\partial f_1}{\partial x_3} \\[2ex] \dfrac{\partial f_3}{\partial x_0} = -\dfrac{\partial f_2}{\partial x_1} = -\dfrac{\partial f_1}{\partial x_2} = \dfrac{\partial f_0}{\partial x_3}. \end{cases}$$

In this case, a similar fact is true.

Theorem 2 *Let $U \subseteq \mathbb{R}^{2,2}$ be open and connected and $F : U \to C\ell_{1,1}$. Then F is left $C\ell_{1,1}$-differentiable if and only $F(Z) = ZA + K$, where $A, K \in C\ell_{1,1}$. That is, the left $C\ell_{1,1}$-differentiable functions must be affine mappings.*

Remark 6 Thus, right or left $C\ell_{1,1}$-differentiability is, perhaps, not a good analogue of holomorphy. Even though these are equivalent notions in the complex setting, in the split quaternionic setting there are more directions in which to take the limit and this requires much stronger conditions. For this reason we are justified in studying functions in the kernels of the operators, and not the $C\ell_{1,1}$-differentiable functions.

2.3 Regularity and John's Equation

Given a $F : U \to C\ell_{1,1}$ whose components are at least C^2 and which satisfies at least one of the following:

$$\overline{\partial} F = 0, \ F\overline{\partial} = 0, \ \partial F = 0, \ \text{or} \ F\partial = 0,$$

must have components which satisfy John's equation [12]:

$$\Delta_{2,2} u = 0.$$

Such functions are said to be ultra-hyperbolic.

In fact, we can use ultra-hyperbolic functions to build regular functions.

Theorem 3 *Let $f : U \subseteq \mathbb{R}^{2,2} \to C\ell_{1,1}$ ultra-hyperbolic components, then $F = \partial f$ is left regular.*

Proof Clearly,

$$\overline{\partial} F = \Delta_{2,2} f = 0.$$

Remark 7 Since all left and right differentiable functions are affine mappings, their component functions, which are linear combinations of the variables, are ultra-hyperbolic.

2.4 A Brief Discussion of Split-Quaternionic Power Series

In [11], Sudbery introduces a notion of quaternionic power series. Indeed, this idea extends to the split-quaternions. First, note the following, which is mentioned in Sudebry's paper.

Proposition 4 *The components of Z^n are real homogeneous polynomials in the variables x_0, x_1, x_2, x_3 for $n \geq 1$.*

Proof We proceed by induction. The case $n = 1$ is clearly true. So we may assume that for some k that

$$Z^k = P_{k,0} + P_{k,1}i + P_{k,2}j + P_{k,3}ij,$$

where $P_{k,i} = P_{k,i}(x_0, x_1, x_2, x_3)$ is a homogeneous polynomial of degree k. Then a simple calculation shows us that

$$
\begin{aligned}
Z^{k+1} = {} & \left[x_0(P_{k,0}) - x_1(P_{k,1}) + x_2(P_{k,2}) + x_3(P_{k,3}) \right] \\
& + \left[x_0(P_{k,1}) + x_1(P_{k,0}) + x_2(P_{k,3}) - x_3(P_{k,2}) \right] i \\
& + \left[x_0(P_{k,2}) + x_1(P_{k,3}) + x_2(P_{k,0}) - x_3(P_{k,1}) \right] j \\
& + \left[x_0(P_{k,3}) - x_1(P_{k,2}) + x_2(P_{k,1}) + x_3(P_{k,0}) \right] ij.
\end{aligned}
$$

Notice that each component is itself a homogeneous polynomial of degree $k + 1$, as required.

Proposition 5 If $A_n = a_n + b_n i + c_n j + d_n ij \in C\ell_{1,1}$ then the components of $A_n Z^n$ are also homogeneous polynomials of degree n.

Proof By above we know Z^n has components which are homogeneous polynomials. That is,

$$Z^n = P_{n,0} + P_{n,1} i + P_{n,2} j + P_{n,3} ij.$$

But,

$$\begin{aligned}
A_n Z^n = & \left[a_n(P_{n,0}) - b_n(P_{n,1}) + c_n(P_{n,2}) + d_n(P_{n,3}) \right] \\
& + \left[a_n(P_{n,1}) + b_n(P_{n,0}) + c_n(P_{n,3}) - d_n(P_{n,2}) \right] i \\
& + \left[a_n(P_{n,2}) + b_n(P_{n,3}) + c_n(P_{n,0}) - d_n(P_{n,1}) \right] j \\
& + \left[a_n(P_{n,3}) - b_n(P_{n,2}) + c_n(P_{n,1}) + d_n(P_{n,0}) \right] ij,
\end{aligned}$$

and the components are homogeneous polynomials.

Thus, a theory of split-quaternionic power series is the same as the theory of real power series in four variables.

Remark 8 Notice that since AZ, nor Z^2 is annihilated by ∂ or its conjugate that we cannot expect a function which can be expressed as a split-quaternionic power series to be annihilated by the operators either.

3 A Theory of Left-Regular Functions

With all of these notions of holomorphic functions, it becomes necessary to choose one and deem it the "canonical" one. Since the difference quotients do not yield an extensive class of functions, we believe use of an operator to be the be the best place to start. Given the association between $C\ell_{1,1}$ and $\mathbb{R}^{2,2}$, it seems $\overline{\partial}$ is the ideal operator for our purposes, since it is also the gradient in $\mathbb{R}^{2,2}$ (and since it is an analogue of $\partial_{\bar{z}}$, which is $\frac{1}{2}$ times the gradient of $\mathbb{R}^2$). Given the argument in Remark 4, we choose left-regularity to be the canonical notion of holomorphy.

Remark 9 It should be noted that Sudbery reached the same conclusions in [11], and that these ideas have been extended to find analogues of "derivative" in this context for the quaternions, which are called *hyper-derivatives* (see [24]). It is conceivable that such results would have some analogue in the split-quaternions.

It should also be noted that left-regularity is the analogue of holomorphy chosen by Libine [12]. In his work, he shows that left-regular functions satisfy a Cauchy-like integral formula.

Theorem 4 *(Libine's Integral Formula) Let $U \subseteq C\ell_{1,1}$ be a bounded open (in the Euclidean topology) region with smooth boundary ∂U. Let $f : U \to C\ell_{1,1}$ be a function which extends to a real-differentiable function on an open neighborhood $V \subseteq C\ell_{1,1}$ of $\overline{U}$ such that $\overline{\partial} f = 0$. Then for any $Z_0 \in C\ell_{1,1}$ such that the boundary of U intersects the cone $C = \left\{ Z \in C\ell_{1,1} : (Z - Z_0)\overline{(Z - Z_0)} = 0 \right\}$ transversally, we have*

$$\lim_{\epsilon \to 0} \frac{-1}{2\pi^2} \int_{\partial U} \frac{\overline{(Z - Z_0)}}{(Z - Z_0)\overline{(Z - Z_0)} + i\epsilon \, \|Z - Z_0\|^2} \cdot dZ \cdot f(Z)$$

$$= \begin{cases} f(Z_0) & \text{if } Z_0 \in U \\ 0 & \text{else} \end{cases},$$

where the three form dZ is given by

$$dZ = dx_1 \wedge dx_2 \wedge dx_3 - (dx_0 \wedge dx_2 \wedge dx_3)i + (dx_0 \wedge dx_1 \wedge dx_3)j - (dx_0 \wedge dx_1 \wedge dx_2)ij.$$

Given this interesting property has been proven, it is some-what surprising that a more detailed description of left-regular functions has not been given in the literature. So we conclude by showing that some left regular functions have a simple description.

3.1 A Class of Left Regular Functions

To date, the author has not been able to find a description for left regular functions in a manner similar to the split-complex case [14, 20]. It may be the case that no such description exists in general. However, it is possible to give a large class of left-regular functions a simple description.

Theorem 5 *Let $F : U \subseteq \mathbb{R}^{2,2} \to C\ell_{1,1}$ have the form*

$$\begin{aligned} F(x_0, x_1, x_2, x_3) &= (g_1(x_0 + x_2, x_1 + x_3) + g_2(x_0 - x_2, x_1 - x_3)) \\ &+ (g_3(x_0 - x_2, x_1 - x_3) + g_4(x_0 + x_2, x_1 + x_3)) \, i \\ &+ (g_1(x_0 + x_2, x_1 + x_3) - g_2(x_0 - x_2, x_1 - x_3)) \, j \\ &+ (g_3(x_0 - x_2, x_1 - x_3) - g_4(x_0 + x_2, x_1 + x_3)) \, ij, \end{aligned}$$

where $g_i \in C^1(U)$. Then $\overline{\partial} F = 0$.

Proof We can easily check that such an F satisfies the necessary system of PDEs. However, it is far more enlightening to see how one can arrive at such a solution.

Write $F = f_0 + f_1 i + f_2 j + f_3 ij$. Using an argument from [14], we have that

$$\overline{\partial} = \left(\frac{\partial}{\partial x_0} - j\frac{\partial}{\partial x_2}\right) + i\left(\frac{\partial}{\partial x_1} - j\frac{\partial}{\partial x_3}\right)$$

$$= 2\left(\frac{\partial}{\partial v_0}j_+ + \frac{\partial}{\partial u_0}j_-\right) + 2i\left(\frac{\partial}{\partial v_1}j_+ + \frac{\partial}{\partial u_1}j_-\right)$$

$$:= \partial_1 + i\partial_2,$$

where $u_0 = x_0 + x_2$, $v_0 = x_0 - x_2$, $u_1 = x_1 + x_3$, $u_1 = x_1 - x_3$, $j_+ = \dfrac{1+j}{2}$, and $j_- = \dfrac{1-j}{2}$.

The key fact is that j_+ and j_- are idempotents and annihilate each other. Also, notice that $ij_+ = j_- i$ and $ij_- = j_+ i$.

Similarly, we may write

$$F = (F_0 j_+ + F_1 j_-) + i(F_2 j_+ + F_3 j_-).$$

Now, *one way* in which $\overline{\partial}F = 0$ is if

$$\partial_1(F_0 j_+ + F_1 j_-) = \partial_2(F_0 j_+ + F_1 j_-)$$

$$= \partial_1(i(F_2 j_+ + F_3 j_-)) = \partial_2(i(F_2 j_+ + F_3 j_-)) = 0.$$

Using the above facts about j_+ and j_-, we see that the conditions implies that

$$\begin{cases} \dfrac{\partial F_0}{\partial v_0} = \dfrac{\partial F_0}{\partial v_1} = 0 \\[2mm] \dfrac{\partial F_1}{\partial u_0} = \dfrac{\partial F_1}{\partial u_1} = 0 \\[2mm] \dfrac{\partial F_2}{\partial u_0} = \dfrac{\partial F_2}{\partial u_1} = 0 \\[2mm] \dfrac{\partial F_3}{\partial v_0} = \dfrac{\partial F_3}{\partial v_1} = 0 \end{cases}$$

This, of course, means that

$$F_0 = F_0(u_0, u_1), \quad F_1 = F_1(v_0, v_1), \quad F_2 = F_2(v_0, v_1), \quad F_3 = F_3(u_0, u_1).$$

Translating back to the original coordinates, we see F has the desired form.

The converse is not true, in general. Here is a simple counter-example.

Example 4 Consider the $C\ell_{1,1}$-valued function

$$f(x_0, x_1, x_2, x_3) = x_1 x_2 x_3 - x_0 x_2 x_3 i + x_0 x_1 x_3 j + x_0 x_1 x_2 i j.$$

It is easy to check that f satisfies the necessary system of PDEs so that $\overline{\partial} f = 0$. However, notice that if we write f as in the above proof, then

$$f = \frac{\left(u_1^2 - v_1^2\right)}{4}\left(u_0 j_+ - v_0 j_-\right) + i\frac{\left(u_0^2 v_0^2\right)}{4}\left(-u_0 j_+ + v_0 j_-\right).$$

Now,

$$\partial_2 \left[\frac{\left(u_1^2 - v_1^2\right)}{4}\left(u_0 j_+ - v_0 j_-\right)\right] = -2v_1 u_0 j_+ - 2u_1 v_0 \neq 0.$$

Thus, f is not of the form as prescribed in Theorem 5.

3.2   Generating Left Regular Functions

In a manner similar to the $C\ell_{0,n}$ case, we can also take a $C\ell_{1,1}$-valued function whose components are real analytic and generate a left regular function valued in $C\ell_{1,1}$. In fact, there are two ways to do this. The first borrows heavily from a result found in Brackx, Delanghe, and Sommen's book [9].

Theorem 6 *Let $g(x_2, x_3)$ be a $C\ell_{1,1}$-valued function on $U \subseteq \mathbb{R}^2$ with real-analytic components. Then the function*

$$f(Z) = \sum_{k=0}^{\infty} \partial \left[\left(\frac{x_0^{2k+1} + x_1^{2k+1}}{(2k+1)!}\right) \Delta^k g(x_2, x_3)\right],$$

where Δ is the Laplace operator in the $x_2 x_3$-plane, is left-regular in an open neighborhood of $\{(0,0)\} \times U$ in $\mathbb{R}^{2,2}$ and $f(0, 0, x_2, x_3) = g(x_2, x_3) - i g(x_2, x_3)$.

Proof (Proof of the Theorem) We proceed by a similar proof found in [9].

Let $g(x_2, x_3) = g_0(x_2, x_3) + g_1(x_2, x_3)i + g_2(x_2, x_3)j + g_3(x_2, x_3)ij$. Since g_ℓ is analytic, then an application of Taylor's theorem gives that on every compact set $K \subset U$ there are constants c_K and λ_K, depending on K, such that

$$\sup_{(x_2,x_3)\in K} \left|\Delta^k g(x_2, x_3)\right| \leq (2k)! c_K \lambda_K^k \quad \text{and}$$

$$\sup_{(x_2,x_3)\in K} \left|\frac{\partial}{\partial x_\ell}\Delta^k g(x_2, x_3)\right| \leq (2k+1)! c_K \lambda_K^k,$$

where $|\cdot|$ denotes the euclidean norm in $\mathbb{R}^4$.

Thus,

$$
\sup_{(x_2,x_3)\in K} \left| \partial \left[\left(\frac{x_0^{2k+1} + x_1^{2k+1}}{(2k+1)!} \right) \Delta^k g(x_2, x_3) \right] \right|
$$

$$
= \sup_{(x_2,x_3)\in K} \left| \left(\frac{x_0^{2k}}{(2k)!} + i\frac{x_1^{2k}}{(2k)!} \right) \Delta^k g \right.
$$

$$
\left. - \frac{x_0^{2k+1} + x_1^{2k+1}}{(2k+1)!} \left(j\frac{\partial}{\partial x_2} \Delta^k g + ij\frac{\partial}{\partial x_3} \Delta^k g \right) \right|
$$

$$
\leq \sup_{(x_0,x_1)\in K} \left[\left| \frac{x_0^{2k}}{(2k)!} \right| \left| \Delta^k g \right| + \left| \frac{x_1^{2k}}{(2k)!} \right| \left| \Delta^k g \right| \right.
$$

$$
\left. + \left| \frac{x_0^{2k+1} + x_1^{2k+1}}{(2k+1)!} \right| \left(\left| \frac{\partial}{\partial x_2} \Delta^k g \right| + \left| \frac{\partial}{\partial x_3} \Delta^k g \right| \right) \right]
$$

$$
\leq c_K \left[(1 + 2|x_0|) x_0^{2k} \lambda_K^k + (1 + 2|x_1|) x_1^{2k} \lambda_K^k \right],
$$

so that f converges uniformly on

$$
\bigcup_{K \subseteq U} \left[\left(-\frac{1}{\sqrt{\lambda_K}}, \frac{1}{\sqrt{\lambda_K}} \right) \times \left(-\frac{1}{\sqrt{\lambda_K}}, \frac{1}{\sqrt{\lambda_K}} \right) \times \mathring{K} \right].
$$

Now,

$$
\bar{\partial} f = \sum_{k=0}^{\infty} \Delta_{2,2} \left[\left(\frac{x_0^{2k+1} + x_1^{2k+1}}{(2k+1)!} \right) \Delta^k g(x_2, x_3) \right]
$$

$$
= \sum_{k=0}^{\infty} \Delta_{2,2} \left(\frac{x_0^{2k+1} + x_1^{2k+1}}{(2k+1)!} \right) \Delta^k g(x_2, x_3)
$$

$$
+ \sum_{k=0}^{\infty} \left(\frac{x_0^{2k+1} + x_1^{2k+1}}{(2k+1)!} \right) \Delta_{2,2} \left(\Delta^k g(x_2, x_3) \right)
$$

$$
= \sum_{k=1}^{\infty} \left(\frac{x_0^{2k-1} + x_1^{2k-1}}{(2k-1)!} \right) \Delta^k g(x_2, x_3)
$$

$$
- \sum_{k=0}^{\infty} \left(\frac{x_0^{2k+1} + x_1^{2k+1}}{(2k+1)!} \right) \Delta^{k+1} g(x_2, x_3)
$$

$$
= 0,
$$

as needed.

Remark 10 Notice that without the ∂ in the series above, one actually produces a $C\ell_{1,1}$-valued function that is ultra-hyperbolic.

Example 5 Let $g(x_2, x_3) = x_2 x_3$. Then $\Delta g = 0$ and the formula above gives

$$f(Z) = \partial \left[(x_0 + x_1)(x_2 x_3) \right]$$
$$= x_2 x_3 - x_2 x_3 i + (x_0 x_3 + x_1 x_3) j + (x_0 x_2 + x_1 x_2).$$

A less trivial example demonstrates that the more complicated g is the more complicated f is.

Example 6 Let $g(x_2, x_3) = x_2^4 + x_2 x_3^3$. Thus, $\Delta g = 12x_2^2 + 6x_2 x_3$ and $\Delta^2 g = 24$. Then from the formula, we get

$$f(Z) = \left[x_0^4 + 3x_0^2 (2x_2^2 + x_2 x_3) + x_2^4 + x_2 x_3^3 \right]$$
$$+ \left[x_1^4 + 3x_1^2 (2x_2^2 + x_2 x_3) + x_2^4 + x_2 x_3^3 \right] i$$
$$+ \left[(x_0^2 x_1^2) + x_2^2 (x_0^2 + x_1^2) + \frac{x_2^4}{3} \right] j$$
$$+ \left[(x_0^2 x_1^2) + x_3^2 (x_0^2 + x_1^2) + \frac{x_3^4}{3} \right] ij.$$

We can define a true extension of an analytic function which is left regular and closely resembles the Cauchy-Kowalewski extension found in [8, 10]. Again, we are again grateful to Brackx et. al for their proof in the $C\ell_{0,n}$ case, which again gives the convergence of the series.

Theorem 7 *(Cauchy-Kowalewski Extension in $C\ell_{1,1}$) Let $g(x_1, x_2, x_3)$ be a $C\ell_{1,1}$-valued function whose components are real-analytic functions on $U \subseteq \mathbb{R}^3$. Then the function*

$$f(x_0, x_1, x_2, x_3) = \sum_{k=0}^{\infty} \frac{(-x_0)^k}{k!} D^k g(x_1, x_2, x_3),$$

where $D = \overline{\partial} - \frac{\partial}{\partial x_0}$, is left-regular in an open neighborhood of $\{0\} \times U$, and $f(0, x_1, x_2, x_3) = g(x_1, x_2, x_3)$. Further, f is the unique function with these properties.

The following lemma will be useful in demonstrating the convergence of f in an open neighborhood of U.

Lemma 2 *Let $g(x_1, x_2, x_3)$ be a $C\ell_{1,1}$-valued function whose components are real-analytic functions on $U \subseteq \mathbb{R}^3$. Then on a compact set K, there are constants c_K and λ_K such that*

$$\left| D^k g(x_1, x_2, x_3) \right| \le 3^k c_K(k!)\lambda_K^k.$$

Proof (Proof of the Lemma) Taylor's theorem, again, gives that on a compact set K, there are constants c_K and λ_K such that

$$\left| \frac{\partial^k}{\partial x_1^{k_1} \partial x_2^{k_2} \partial x_3^{k_3}} g(x_1, x_2, x_3) \right| \le c_K(k!)\lambda_K^k.$$

It is worth mentioning that we first saw the above inequality in [9].

Notice that when k is even, D^k is a scalar operator. So suppose k is even. Then by the trinomial theorem,

$$D^k = \sum_{k_1+k_2+k_3=k} \frac{k!}{(k_1)!(k_2)!(k_3)!} \frac{\partial^k}{\partial x_1^{k_1} \partial x_2^{k_2} \partial x_3^{k_3}}.$$

Then,

$$\left| D^k g(x_1, x_2, x_3) \right| \le \sum_{k_1+k_2+k_3=k} \frac{k!}{(k_1)!(k_2)!(k_3)!} \left| \frac{\partial^k}{\partial x_1^{k_1} \partial x_2^{k_2} \partial x_3^{k_3}} g(x_1, x_2, x_3) \right|$$

$$\le c_K(k!)\lambda_K \sum_{k_1+k_2+k_3=k} \frac{k!}{(k_1)!(k_2)!(k_3)!}$$

$$= 3^k c_K(k!)\lambda_K^k.$$

Now suppose k is odd. Then we have

$$D^k = \sum_{\substack{k_1+k_2+k_3= \\ k-1}} \frac{(k-1)!}{(k_1)!(k_2)!(k_3)!} \left(i \frac{\partial^k}{\partial x_1^{k_1+1} \partial x_2^{k_2} \partial x_3^{k_3}} \right.$$

$$\left. - j \frac{\partial^k}{\partial x_1^{k_1} \partial x_2^{k_2+1} \partial x_3^{k_3}} - ij \frac{\partial^k}{\partial x_1^{k_1} \partial x_2^{k_2} \partial x_3^{k_3+1}} \right).$$

This means that

$$\left| D^k g(x_1, x_2, x_3) \right| \leq \sum_{\substack{k_1+k_2+k_3= \\ k-1}} \frac{(k-1)!}{(k_1)!(k_2)!(k_3)!} \left| i \frac{\partial^k g(x_1, x_2, x_3)}{\partial x_1^{k_1+1} \partial x_2^{k_2} \partial x_3^{k_3}} \right.$$

$$\left. - j \frac{\partial^k g(x_1, x_2, x_3)}{\partial x_1^{k_1} \partial x_2^{k_2+1} \partial x_3^{k_3}} - ij \frac{\partial^k g(x_1, x_2, x_3)}{\partial x_1^{k_1} \partial x_2^{k_2} \partial x_3^{k_3+1}} \right|$$

$$\leq \sum_{\substack{k_1+k_2+k_3= \\ k-1}} \frac{(k-1)!}{(k_1)!(k_2)!(k_3)!} \left(\left| \frac{\partial^k g(x_1, x_2, x_3)}{\partial x_1^{k_1+1} \partial x_2^{k_2} \partial x_3^{k_3}} \right| \right.$$

$$\left. + \left| \frac{\partial^k g(x_1, x_2, x_3)}{\partial x_1^{k_1} \partial x_2^{k_2+1} \partial x_3^{k_3}} \right| + \left| \frac{\partial^k g(x_1, x_2, x_3)}{\partial x_1^{k_1} \partial x_2^{k_2} \partial x_3^{k_3+1}} \right| \right)$$

$$= \sum_{\substack{k_1+k_2+k_3= \\ k-1}} \frac{(k-1)!}{(k_1)!(k_2)!(k_3)!} \left(3c_K(k!)\lambda_K^k \right)$$

$$= 3^k c_K(k!)\lambda_K^k,$$

as required.

Proof (Proof of the Theorem) The above lemma gives us that on a compact set $K \subset U$ there are constants c_K and λ_K, depending on K, such that

$$\left| \frac{(-x_0)^k}{k!} D^k g(x_1, x_2, x_3) \right| \leq c_K (3\lambda_K)^k x_0^k.$$

Thus, f converges uniformly on

$$\bigcup_{K \subseteq U} \left[\left(-\frac{1}{3\lambda_K}, \frac{1}{3\lambda_K} \right) \times \mathring{K} \right].$$

The essential calculation is

$$\bar{\partial} f = \sum_{k=0}^{\infty} \bar{\partial} \left(\frac{(-x_0)^k}{k!} D^k g(x_1, x_2, x_3) \right)$$

$$= \sum_{k=0}^{\infty} \bar{\partial} \left(\frac{(-x_0)^k}{k!} \right) D^k g(x_1, x_2, x_3) + \sum_{k=0}^{\infty} \frac{(-x_0)^k}{k!} \bar{\partial} \left(D^k g(x_1, x_2, x_3) \right)$$

$$= -\sum_{k=1}^{\infty} \frac{(-x_0)^{k-1}}{(k-1)!} D^k g(x_1, x_2, x_3) + \sum_{k=0}^{\infty} \frac{(-x_0)^k}{k!} D^{k+1} g(x_1, x_2, x_3)$$

$$= 0,$$

as required.

The uniqueness of f follows directly from the arguments raised in [9].

Remark 11 We may think of the above extension as a solution to the boundary value problem:

$$\begin{cases} \overline{\partial} f(x_0, x_1, x_2, x_3) = 0 \\ f(0, x_1, x_2, x_3) = g(x_1, x_2, x_3) \end{cases}.$$

Example 7 Consider the homogeneous polynomial of degree 2

$$g(x_1, x_2, x_3) = x_1^2 + x_2^2 + x_3^2 + x_1 x_2 + x_1 x_3 + x_2 x_3.$$

Now,

$$Dg(x_1, x_2, x_3) = (2x_1 + x_2 + x_3) i - (2x_2 + x_1 + x_3) j - (2x_3 + x_2 + x_1) ij$$

$$D^2 g(x_1, x_2, x_3) = -6$$

$$D^3(x_1, x_2, x_3) = 0.$$

Thus,

$$f(Z) = \left(-3x_0^2 + x_1^2 + x_2^2 + x_3^2 + x_1 x_2 + x_1 x_3 + x_2 x_3\right)$$
$$- x_0 (2x_1 + x_2 + x_3) i - x_0 (2x_2 + x_1 + x_3) j - x_0 (2x_3 + x_1 + x_2) ij.$$

is the left- regular function obtained by the above theorem.

Remark 12 In the formulas of both extensions, a polynomial g will be transformed to a Clifford valued function where every component is a polynomial. This is the case because polynomials have partial derivatives of 0 after a certain order. That is, $D^k g$ and $\Delta^k g$ will be uniformly 0 for all $k > M$ for some finite M.

References

1. Ahlfors, L.A.: Complex Analysis: An Introduction to the Theory of Analytic Functions of One Complex Variable. McGraw-Hill, New York (1979)
2. Jones, G.A., Singerman, D.: Complex Functions: An Algebraic and Geometric Viewpoint. Cambridge University Press (1987)
3. Gilbert, J.E., Murray, M.A.E.: Clifford Algebras and Dirac Operators in Harmonic Analysis. Cambridge University Press (1991)
4. Altmann, S.L.: Hamilton, rodrigues, and the quaternion scandal. Mat. Mag. **62**(5), 291–308 (1989)
5. Graves, R.P.: Life of Sir William Rowan Hamilton. Vol. 3. Hodges, Figgis, and Co., Dublin (1882–1899)
6. Clifford, W.K.: Applications of Grassmann's extensive algebra. Am. J. Math. **1**(4), 350–358 (1878)
7. Dirac, P.A.M.: The quantum theory of the electron. Proc. Royal Soc. Lond. Ser. A Contain. Papers Math. Phys. Character **117**(778), 610–624 (1928)
8. Delanghe, R.: Clifford analysis: history and perspective. Comput. Methods Funct. Theor. **1**(1), 107–153 (2001)

9. Brackx, F., Delanghe, R., Sommen, F.: Clifford Analysis. Pitman Andvanced Publishing Program, Boston (1982)
10. Ryan, J.: Introductory clifford analysis. In: Ablamowicz, R., Sobczyk, G. (eds) Lectures on Clifford(Geometric) Algebras and Applications. Birkhäuser Boston (2004)
11. Sudbery, A.: Quaternionic analysis. Math. Proc. Camb. Philos. Soc. **85**(2), 199–224 (1979)
12. Libine, M.: An Invitation to split quaternionic analysis. In: Sabadini, I., Sommen, F. (eds) Hypercomplex Analysis and Applications, Trends in Mathematics, pp. 161–180. Springer Basel (2011)
13. Masrouri, N., Yayli, Y., Faroughi, M.H., Mirshafizadeh, M.: Comments on differentiable over function of split quaternions. Revista Notas de Matemática **7**(2), 128–134 (2011)
14. Emanuello, J.A., Nolder, C.A.: Projective compactification of $\mathbb{R}^{1,1}$ and its möbius geometry. Complex Anal. Oper. Theor. **9**(2), 329–354 (2015)
15. Kisil, V.V.: Analysis in $\mathbf{R}^{1,1}$ or the principal function theory, complex variables. Theor. Appl. Int. J. **40**(2), 93–118 (1999)
16. Deakin, M.A.B.: Functions of a dual or duo variable. Math. Mag. **39**(4), 215–219 (1966)
17. DenHartigh, K., Flim, R.: Liouville theorems in the dual and double planes. Rose-Hulman Undergrad. Math. J. **12**(2) (2011). Article 4
18. Kisil, Vladimir V.: Two-Dimensional conformal models of space-time and their compactification. J. Math. Phys. **48**(7) (2007). https://doi.org/10.1063/1.2747722
19. Harkin, A.A., Harkin, J.B.: Geometry of generalized complex numbers. Math. Mag. **77**(2), 118–129 (2004)
20. Libine, M.: Hyperbolic cauchy integral formula for the split complex numbers (2007). ArXiv: 0712.0375
21. O'Neill, B.: Semi-Riemmanian Geometry - with Applications to Relativity, Pure and Applied Matematics, 103. Academic Press Inc., New York (1983)
22. Obolashvili, E.: Partial Differential Equations in Clifford Analysis. Addison Wesley Longman Limited, London (1998)
23. Frenkel, I., Libine, M.: Split quaternionic analysis and the separation series for $SL(2, \mathbb{R})$. Adv. Math. **220**, 678–763 (2011)
24. Luna-Elizarrarás, M.E., Shapiro, M.: A survey on the (hyper-) derivatives in complex. Quaternionic Clifford Anal. Milan J. Math. **79**, 521–542 (2011)

Decomposition of the Twisted Dirac Operator

T. Raeymaekers

Abstract The classical Dirac operator is a conformally invariant first order differential operator mapping spinor-valued functions to the same space, where the spinor space is to be interpreted as an irreducible representation of the spin group. In this article we twist the Dirac operator by replacing the spinor space with an arbitrary irreducible representation of the spin group. In this way, the operator becomes highly reducible, whence we determine its full decomposition.

Keywords Dirac operator · Higher spin · Clifford analysis

1 Introduction

This article is written in the setting of Clifford analysis, a function theory which on the one hand generalises complex analysis to the case of an arbitrary dimension m. On the other hand, Clifford analysis is a refinement of harmonic analysis in $\mathbb{R}^m$. At the very heart of the theory lies the (massless) Dirac operator ∂_x, which originated from particle physics [8]. The Dirac operator factorises the Laplace operator $\Delta_x = -\partial_x^2$, reflecting this refinement property. We refer the reader to [1, 7, 13] for a general introduction of this branch of analysis.

In a series of recent papers [2, 6, 9], Clifford analysis techniques have been used to develop a function theory for invariant operators acting on functions taking values in general irreducible representations for the spin group instead of the spinor space. In [5], the Dirac operator was twisted in the sense that instead of letting it act on spinor-valued functions, the operator acts on functions taking values in an arbitrary irreducible spin-representation. By doing this, the twisted Dirac operator can be written as a sum of invariant operators, so-called higher spin operators, which were

T. Raeymaekers (✉)
Department of Mathematical Analysis, Ghent University,
Krijgslaan 281, 9000 Ghent, Belgium
e-mail: Tim.Raeymaekers@UGent.be

© Springer Nature Switzerland AG 2018
P. Cerejeiras et al. (eds.), *Clifford Analysis and Related Topics*,
Springer Proceedings in Mathematics & Statistics 260,
https://doi.org/10.1007/978-3-030-00049-3_6

constructed in [9, 18]. The aim of the present paper is to determine this decomposition explicitly, hereby making use of representation theoretical as well as function theoretical results.

After introducing basic notations and definitions in Sect. 2, we define the twisted Dirac operator in Sect. 3, where we also make the theoretical prediction of the decomposition using Stein-Weiss gradients [18]. In Sects. 4 and 5, we calculate the puzzle pieces needed to explicitly determine this decomposition in Sect. 7.

2 Definitions and Notations

Let $(e_1, \ldots, e_m)$ be an orthonormal basis for the m-dimensional Euclidean vector space $\mathbb{R}^m$. The real Clifford algebra $\mathbb{R}_m$ is defined as the algebra generated by these basis elements under the multiplication relations $e_i e_j + e_j e_i = -2\delta_{ij}$, for all $1 \leq i, j \leq m$. As a vector space over $\mathbb{R}$, this algebra has dimension 2^m. The complex Clifford algebra $\mathbb{C}_m$ is the complexification of the real Clifford algebra, $\mathbb{C}_m := \mathbb{R}_m \otimes \mathbb{C}$. Any vector $x = (x_1, \ldots, x_m) \in \mathbb{R}^m$ can be embedded inside the Clifford algebra as $x \hookrightarrow \sum_{j=1}^m e_j x_j$. Throughout the article, we use the same notation x for both objects (the meaning will always be clear from the context). The group $\mathrm{Spin}(m)$ can be realised within the Clifford algebra itself as the set of even products of unit vectors:

$$\mathrm{Spin}(m) = \left\{ s = \prod_{j=1}^{2a} \omega_j : a \in \mathbb{N}, \omega_j \in S^{m-1} \right\},$$

where S^{m-1} denotes the unit sphere in $\mathbb{R}^m$. Then the Dirac spinor space $\mathbb{S}$ can be realised as a minimal left ideal in $\mathbb{C}_m$, see e.g. [7] for the explicit definition. It is crucial to point out that the vector space $\mathbb{S}$ defines a model for a representation of the spin group. For an odd dimension $m = 2n + 1$, $\mathbb{S}$ is the irreducible basic half-integer representation of $\mathrm{Spin}(m)$, labeled with the highest weight $(\frac{1}{2}, \frac{1}{2}, \ldots, \frac{1}{2})$, with n entries. In case of an even dimension $m = 2n$, the spinor space is a reducible representation: $\mathbb{S} = \mathbb{S}^+ \oplus \mathbb{S}^-$, where $\mathbb{S}^+$ and $\mathbb{S}^-$ are irreducible representations for $\mathrm{Spin}(m)$, with highest weights $(\frac{1}{2}, \frac{1}{2}, \ldots, \frac{1}{2})$ and $(\frac{1}{2}, \frac{1}{2}, \ldots, \frac{1}{2}, -\frac{1}{2})$ respectively, the so-called positive and negative Weyl spinors.

The Dirac operator ∂_x is the unique (up to a multiplicative constant) conformally invariant elliptic first order differential operator given by

$$\partial_x = \sum_{j=1}^m e_j \partial_{x_j} : \mathscr{C}^\infty(\mathbb{R}^m, \mathbb{S}) \to \mathscr{C}^\infty(\mathbb{R}^m, \mathbb{S}).$$

In case m is even, this operator reverses spinor parities: $\partial_x : \mathscr{C}^\infty(\mathbb{R}^m, \mathbb{S}^\pm) \rightarrow \mathscr{C}^\infty(\mathbb{R}^m, \mathbb{S}^\mp)$. For notational convenience, we restrict ourselves to the odd-dimensional case, but a completely similar reasoning can be made for the even dimensional case, taking into account this change of parity.

In fact, all irreducible Spin(m)-representations can be modeled in the language of Clifford analysis, as spaces of polynomials in several (dummy) vector variables $u_i \in \mathbb{R}^m$, see e.g. [4, 13]. Throughout this article, we assume that the number of variables $k < \lfloor \frac{m}{2} \rfloor$, thus restricting ourselves to the so-called stable range [14]. As we frequently need the Dirac operators associated to each of the variables $u_i \in \mathbb{R}^m$, we denote the corresponding operator ∂_{u_i} by means of ∂_i. The bracket notation $\langle \cdot, \cdot \rangle$ represents the Euclidean inner product on $\mathbb{R}^m$. As Dirac operators are vector-like objects, we use the Euclidean inner product for them as well, e.g. $\langle \partial_x, \partial_y \rangle = \sum_{j=1}^m \partial_{x_j} \partial_{y_j}$.

Definition 6.1 A function $f : \mathbb{R}^{k \times m} \rightarrow \mathbb{C} : (u_1, \ldots, u_k) \mapsto f(u_1, \ldots, u_k)$ is called simplicial harmonic if it satisfies $\langle \partial_i, \partial_j \rangle f = 0$ for all $1 \le i, j \le m$ and $\langle u_i, \partial_j \rangle f = 0$ for all $i < j$.

The space of $\mathbb{C}$-valued simplicial harmonic polynomials which are homogeneous of degree l_i in u_i will be denoted by $\mathscr{H}_{l_1, \ldots, l_k}$ or $\mathscr{H}_\lambda$ for short, with $\lambda = (l_1, \ldots, l_k)$. $\mathscr{H}_\lambda$ is a model for the irreducible Spin(m)-representation with integer highest weight $(l_1, \ldots, l_k, 0, \ldots, 0)$, which we shall from now on denote by $(l_1, \ldots, l_k)$ or λ for short. Since a highest weight satisfies the dominant weight condition, we will from now on assume that $l_1 \ge \cdots \ge l_k \ge 0$ (as an a priori condition on the degrees of homogeneity).

For the half integer Spin(m)-representations, we introduce the following function space:

Definition 6.2 A function $f : \mathbb{R}^{k \times m} \rightarrow \mathbb{S} : (u_1, \ldots, u_k) \mapsto f(u_1, \ldots, u_k)$ is called simplicial monogenic if it satisfies $\partial_i f = 0$ for all i and $\langle u_i, \partial_j \rangle f = 0$ for all $i < j$.

The space of $\mathbb{S}$-valued simplicial monogenic polynomials which are homogeneous of degree l_i in u_i will be denoted by $\mathscr{S}_{l_1, \ldots, l_k}$ or $\mathscr{S}_\lambda$ for short, with $\lambda = (l_1, \ldots, l_k)$. The space $\mathscr{S}_\lambda$ defines a model for the irreducible Spin(m)-representation with half-integer highest weight $(l_1 + \frac{1}{2}, \ldots, l_k + \frac{1}{2}, \frac{1}{2}, \ldots, \frac{1}{2})$, which we shall from now on denote by $(l_1, \ldots, l_k)'$ or λ' for short.

The polynomial models for spin representations can now be used to define the twisted Dirac operator.

3 The Twisted Dirac Operator

In this section we consider the twisted Dirac operator, which can be defined on $\mathscr{S}_\lambda$-valued functions for arbitrary λ.

Definition 6.3 For arbitrary integer-valued highest weights λ for Spin(m), the twisted Dirac operator on $\mathscr{S}_\lambda$-valued polynomials is defined as

$$\partial_x^T := \mathbf{1}_\lambda \otimes \partial_x : \mathscr{C}^\infty(\mathbb{R}^m, \mathscr{S}_\lambda) \to \mathscr{C}^\infty(\mathbb{R}^m, \mathscr{H}_\lambda \otimes \mathbb{S}).$$

The superindex T represents the twisted nature of the operator.

Remark 6.1 Since ∂_x and ∂_i do not commute, functions in the codomain of ∂_x^T are not necessarily $\mathscr{S}_\lambda$-valued. However, ∂_x^T commutes with all defining operators of $\mathscr{H}_\lambda$, whence $\mathrm{Im}(\partial_x^T)$ must be a subset of $\mathscr{C}^\infty(\mathbb{R}^m, \mathscr{H}_\lambda \otimes \mathbb{S})$;

The value space of functions in the codomain of ∂_x^T is highly reducible, which suggests the reducibility of ∂_x^T. Indeed, we have the following tensor product decomposition.

Lemma 6.1 (e.g. [10]) *As a Spin(m)-representation, the tensor product $\mathscr{H}_\lambda \otimes \mathbb{S}$ can be decomposed as the direct sum of (at most) 2^k irreducible Spin(m)-modules, each one appearing with multiplicity 1:*

$$\mathscr{H}_\lambda \otimes \mathbb{S} \cong \bigoplus_{i_1=0}^{1} \cdots \bigoplus_{i_k=0}^{1} \mathscr{S}_{l_1-i_1,\dots,l_k-i_k}. \tag{1}$$

Each summand $(l_1 - i_1, \dots l_k - i_k)'$ is contained in the decomposition as long as its highest weight satisfies the dominant weight condition.

Note that (1) merely gives an automorphism between spin representations. If we want an equality as function spaces, we need nontrivial embedding factors (operators), embedding an isomorphic copy of a component on the right-hand side into the left-hand side $\mathscr{H}_\lambda \otimes \mathbb{S}$. It is easily seen that these embedding operators are indeed nontrivial in general, as the degrees of homogeneity of the appearing spaces do not match. One of the targets of this paper is to calculate the needed embedding operators (see Sect. 5).

The technique of using generalised gradients in e.g. [11, 18] tells us that not all conformal invariant first order differential operators $\mathscr{C}^\infty(\mathbb{R}^m, \mathscr{S}_\lambda) \to \mathscr{C}^\infty(\mathbb{R}^m, \mathscr{S}_{l_1-i_1,\dots,l_k-i_k})$ are nontrivial. Hence, one can deduce from these results and the lemma above that the twisted Dirac operator $\mathbf{1}_\lambda \otimes \partial_x$ can be written as the sum of at most $k + 1$ first order differential operators: a higher spin Dirac operator $\mathscr{Q}_\lambda$ and (at most) k higher spin twistor operators $\mathscr{T}_\lambda^{(i)}$, defined as the unique conformally invariant first order differential operators mapping $\mathscr{S}_\lambda$-valued functions to $\mathscr{S}_{\lambda-L_i}$-valued functions, where $\lambda - L_i$ stands for $(l_1, \dots, l_i - 1, \dots, l_k)$. Each of these operators is defined through the following scheme:

$$\mathscr{C}^\infty(\mathbb{R}^m, \mathscr{S}_\lambda) \xrightarrow{\ \partial_x^T\ } \mathscr{C}^\infty(\mathbb{R}^m, \mathscr{H}_\lambda \otimes \mathbb{S}) \tag{2}$$

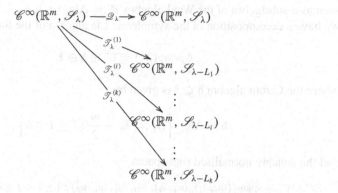

These operators were constructed in [9] and have the following forms:

$$\mathscr{D}_\lambda = \prod_{i=1}^{k}\left(1 + \frac{u_i \partial_i}{m + 2\mathbb{E}_i - 2i}\right)\partial_x, \tag{3}$$

and

$$\mathscr{T}_\lambda^{(j)} = \prod_{p=j+1}^{k}\left(1 - \frac{\langle u_p, \partial_j\rangle\langle u_j, \partial_p\rangle}{\mathbb{E}_j - \mathbb{E}_p + p - j + 1}\right)\langle \partial_j, \partial_x\rangle. \tag{4}$$

As mentioned before, we need embedding operators $\mathscr{E}_i : \mathscr{S}_{\lambda - L_i} \hookrightarrow \mathscr{H}_\lambda \otimes \mathbb{S}$. Since for all $f \in \mathscr{S}_{\lambda - L_i}$, $\mathscr{E}_i f$ has to be simplicial harmonic, we need a so-called extremal projector operator which projects any function on its simplicial harmonic part.

4   An Extremal Projector for the Symplectic Lie Algebra

We will construct this projection operator in a Lie algebraic setting, following the procedure originally developed in [17, 19, 20], and used in the context of Clifford analysis for the Lie superalgebras $\mathfrak{osp}(1, 2k)$ in [9]. We apply the techniques described in the latter article to the particular case of the classical symplectic Lie algebras $\mathfrak{g} = \mathfrak{sp}(2k + 2, \mathbb{C})$ and $\mathfrak{k} = \mathfrak{sp}(2k, \mathbb{C})$. The reason why we choose this Lie algebra is that it can elegantly be modeled in Clifford analysis in k vector variables $(u_1, \ldots, u_k) \in \mathbb{R}^{k \times m}$ and their corresponding Dirac operators $(\partial_1, \ldots, \partial_k)$:

$$\mathfrak{sp}(2k) = \mathrm{Span}_\mathbb{C}\left\{-\frac{1}{2}\Delta_a, -\langle \partial_a, \partial_b\rangle, \langle u_i, \partial_j\rangle,\right.$$

$$\frac{1}{2}|u_a|^2, \langle u_a, u_b\rangle, \langle u_j, \partial_i\rangle : 1 \le i < j \le k, 1 \le a \ne b \le k \Big\},$$

seen as a subalgebra of the Weyl algebra $\mathscr{W} = \mathrm{Alg}_{\mathbb{C}}\{u_{i,j}, \partial_{u_{i,j}}\}$. On the other hand, we have a decomposition of the symplectic Lie algebra of the form

$$\mathfrak{k} = \mathfrak{sp}(2k) = \mathfrak{k}^+ \oplus \mathfrak{h} \oplus \mathfrak{k}^-, \tag{5}$$

where the Cartan algebra $\mathfrak{h} \subset \mathfrak{k}$ is given by

$$\mathfrak{h} = \mathrm{Alg}_{\mathbb{C}} \Big\{ H_i := \mathbb{E}_i + \frac{m}{2} : 1 \le i \le k \Big\},$$

and the suitably normalised root spaces

$$\mathfrak{k}^+ \cup \mathfrak{k}^- := \mathrm{span}_{\mathbb{C}} \left(\langle u_i, \partial_j\rangle, \langle u_j, \partial_i\rangle, \langle \partial_a, \partial_b\rangle, \langle u_a, u_b\rangle : 1 \le i < j \le k, 1 \le a, b \le k \right).$$

A detailed explanation of root spaces can be found in e.g. [12]. Next, we separate the positive root vectors $\mathfrak{k}^+$ from the negative root vectors $\mathfrak{k}^-$ by choosing a suitable functional ℓ on $\mathfrak{h}^*$ which fixes the parity of our roots. Since an extremal projector operator has the property that it projects on the joint kernel of all positive root vectors, we demand our positive root vectors to be precisely the operators defining the simplicial harmonics (see Definition 6.1). Let us choose k real numbers $c_1, c_2, \ldots, c_k$ such that $c_k < \cdots < c_2 < c_1 < 0$, and consider the linear functional

$$\ell(a_1 L_1 + a_2 L_2 + \cdots + a_k L_k) := a_1 c_1 + a_2 c_2 + \cdots + a_k c_k,$$

where we have the standard dual basis $L_i = H_i^*$ for which $L_i(H_j) = \delta_{ij}$ (see e.g. [12]). A root α is considered positive if $\ell(\alpha) > 0$ and negative if $\ell(\alpha) < 0$. Hence

$$\mathfrak{k}^+ = \mathrm{span}_{\mathbb{C}} \Big\{ -\frac{1}{2}\Delta_a, -\langle \partial_a, \partial_b\rangle, \langle u_i, \partial_j\rangle : 1 \le i < j \le k, 1 \le a \ne b \le k \Big\},$$

$$\mathfrak{k}^- = \mathrm{span}_{\mathbb{C}} \Big\{ \frac{1}{2}|u_a|^2, \langle u_a, u_b\rangle, \langle u_j, \partial_i\rangle : 1 \le i < j \le k, 1 \le a \ne b \le k \Big\},$$

whence the set of positive roots is given by $\Delta^+ = \{-2L_a, -L_a - L_b, L_i - L_j : 1 \le i < j \le k, 1 \le a \ne b \le k\}$ and the set of negative roots by $\Delta^- = \{2L_a, L_a + L_b, L_j - L_i : 1 \le i < j \le k, 1 \le a \ne b \le k\}$. With each of these roots α corresponds a root vector e_α according to the following table:

α	$-2L_a$	$-L_a - L_b$	$L_i - L_j$	$2L_a$	$L_a + L_b$	$L_j - L_i$
e_α	$-\frac{1}{2}\Delta_a$	$-\langle \partial_a, \partial_b\rangle$	$\langle u_i, \partial_j\rangle$	$\frac{1}{2}|u_a|^2$	$\langle u_a, u_b\rangle$	$\langle u_j, \partial_i\rangle$

Next, we define for each positive root $\alpha \in \Delta^+$ the corresponding Cartan element h_α by taking the Lie bracket of the root vector e_α and $e_{-\alpha}$:

$$h_{-2L_a} = \left[-\frac{\Delta_a}{2}, \frac{|u_a|^2}{2} \right] = - \left(\mathbb{E}_a + \frac{m}{2} \right)$$

$$h_{-L_a-L_b} = [-\langle \partial_a, \partial_b \rangle, \langle u_a, u_b \rangle] = -(m + \mathbb{E}_a + \mathbb{E}_b)$$

$$h_{L_i-L_j} = [\langle u_i, \partial_j \rangle, \langle u_j, \partial_i \rangle] = \mathbb{E}_i - \mathbb{E}_j.$$

All root vectors must be normalised such that $\forall \alpha \in \Delta^+$, $[h_\alpha, e_\alpha] = 2e_\alpha$, which explains the numerical coefficients and minus signs in our original choices (see above). Define $\rho \in \mathfrak{h}^*$ as half the sum of the positive roots. Then for each $\alpha \in \Delta^+$, we can calculate $\rho(h_\alpha)$:

α	h_α	$\rho(h_\alpha)$
$-2L_a$	$-\mathbb{E}_a - \frac{m}{2}$	a
$-L_a - L_b$	$-\mathbb{E}_a - \mathbb{E}_b - m$	$a + b$
$L_i - L_j$	$\mathbb{E}_i - \mathbb{E}_j$	$j - i$.

Then for each positive root α, we calculate the operator p_α defined as

$$p_\alpha = 1 + \sum_{j=1}^{\infty} \frac{(-1)^j}{j!} \frac{e_{-\alpha}^j e_\alpha^j}{(\phi_\alpha + 1)(\phi_\alpha + 2) \cdots (\phi_\alpha + j)},$$

with $\phi_\alpha = h_\alpha + \rho(h_\alpha)$. The operators corresponding to the roots then are given by

$$p_{-2L_a} = \sum_{s=0}^{\infty} \frac{1}{4^s s!} \frac{\Gamma(-\mathbb{E}_a - \frac{m}{2} + a + 1)}{\Gamma(-\mathbb{E}_a - \frac{m}{2} + a + 1 + s)} |u_a|^{2s} \Delta_a^s,$$

$$p_{-L_a-L_b} = \sum_{s=0}^{\infty} \frac{1}{s!} \frac{\Gamma(-\mathbb{E}_a - \mathbb{E}_b - m + a + b + 1)}{\Gamma(-\mathbb{E}_a - \mathbb{E}_b - m + a + b + s + 1)} \langle u_a, u_b \rangle^s \langle \partial_a, \partial_b \rangle^s,$$

$$p_{L_i-L_j} = \sum_{s=0}^{\infty} \frac{(-1)^s}{s!} \frac{\Gamma(\mathbb{E}_i - \mathbb{E}_j + j - i + 1)}{\Gamma(\mathbb{E}_i - \mathbb{E}_j + j - i + 1 + s)} \langle u_j, \partial_i \rangle^s \langle u_i, \partial_j \rangle^s. \tag{6}$$

Definition 6.4 An ordering of a set of positive roots for a Lie algebra is called normal if any composite root lies between its components.

In order to construct the extremal projector for $\mathfrak{sp}(2k, \mathbb{C})$, we then need to fix a normal ordering on the set of positive roots. For instance, we can consider the normal ordering

$$L_1 - L_2, L_1 - L_3, \ldots, L_1 - L_k, L_2 - L_3, L_2 - L_4, \ldots, L_{k-1} - L_k,$$

$$- 2L_k, \ldots, -2L_3, -L_2 - L_3, -L_1 - L_3, -2L_2, -L_1 - L_2, -2L_1. \quad (7)$$

In view of our explicit model for $\mathfrak{sp}(2k, \mathbb{C})$ in terms of Dirac operators and vector variables, taking the product of the operators defined in (6) according to the normal ordering 7 gives an operator which projects an arbitrary (homogeneous and $\mathbb{C}$-valued) polynomial $P(u_1, \ldots, u_k)$ onto its simplicial harmonic part.

5   Embedding Factors

Recall the twisted Dirac operator introduced in Sect. 3:

$$\partial_x^T : \mathscr{C}^\infty(\mathbb{R}^m, \mathscr{S}_\lambda) \to \mathscr{C}^\infty(\mathbb{R}^m, \mathscr{H}_\lambda \otimes \mathbb{S}).$$

The full tensor product decomposition of $\mathscr{H}_\lambda \otimes \mathbb{S}$ was given earlier in 1. This means that we can decompose the twisted Dirac operator into at most 2^k first-order differential suboperators, by projecting onto each of the subspaces. However, due to [18], only $k + 1$ of them are non-trivial: the higher spin Dirac operator, and k higher spin twistor operators. However, one should be very careful with this *decomposition of the Dirac operator*, since the image spaces of the twistor operators obtained are $\mathscr{C}^\infty(\mathbb{R}^m, \mathscr{S}_{\lambda-L_i}), i = 1, \ldots, k$. Since the codomain of the twisted Dirac operator is $\mathscr{C}^\infty(\mathbb{R}^m, \mathscr{H}_\lambda \otimes \mathbb{S})$, we need non-trivial embedding factors $\mathscr{E}_i$:

$$\mathscr{E}_i : \mathscr{C}^\infty(\mathbb{R}^m, \mathscr{S}_{\lambda-L_i}) \to \mathscr{C}^\infty(\mathbb{R}^m, \mathscr{H}_\lambda \otimes \mathbb{S}).$$

Essentially, these embedding factors have to raise the degree of homogeneity in the vector variable u_i by one. The logical embedding factor would then be $\mathscr{E}_i = u_i$. However, for each $f \in \mathscr{C}^\infty(\mathbb{R}^m, \mathscr{S}_{\lambda-L_i})$, we have that $\langle u_j, \partial_i \rangle u_i f = u_j f \neq 0$, for $j < i$, so we need to project on the space $\mathscr{C}^\infty(\mathbb{R}^m, \mathscr{H}_\lambda \otimes \mathbb{S})$. This exactly is what the extremal projector of $\mathfrak{sp}(2k)$, constructed in the previous section, does. If we choose the normal ordering 7, the embedding operators are of the form

$$\mathscr{E}_i = p_{L_1-L_2} p_{L_1-L_3} \cdots p_{L_1-L_k} p_{L_2-L_3} p_{L_2-L_4} \cdots p_{L_{k-1}-L_k}$$

$$p_{-2L_k} \cdots p_{-2L_3} p_{-L_2-L_3} p_{-L_1-L_3} p_{-2L_2} p_{-L_1-L_2} p_{-2L_1} u_i.$$

This expression can be simplified. Take $f \in \mathscr{C}^\infty(\mathbb{R}^m, \mathscr{S}_{\lambda-L_i})$. We then have the relations

- $\Delta_i u_i f = -\partial_i(-m - 2\mathbb{E}_i - u_i \partial_i)f = 0$,
- $\Delta_j u_i f = 0$ for all $i \neq j$,
- $\langle \partial_i, \partial_j \rangle u_i f = \partial_j f + u_i \langle \partial_i, \partial_j \rangle f = 0$ for all $i \neq j$,
- $\langle \partial_j, \partial_l \rangle u_i f = 0$, for all $i \neq j$ and $i \neq l$,
- $\langle u_a, \partial_b \rangle u_i f = 0$ for all $b \neq i$ and $a < b$.

Thus, the expressions of the embedding factors reduce to

$$\mathscr{E}_i f = p_{L_{k-1}-L_k} \cdots p_{L_2-L_4} p_{L_2-L_3} p_{L_1-L_k} \cdots p_{L_1-L_3} p_{L_1-L_2} u_i f$$
$$= p_{L_{k-1}-L_k} \cdots p_{L_2-L_4} p_{L_2-L_3} p_{L_1-L_k} \cdots p_{L_1-L_i} u_i f.$$

For all $i > 1$, we have that

$$\langle u_1, \partial_i \rangle u_i f = u_1 f \neq 0 \quad \text{and} \quad \langle u_1, \partial_i \rangle^2 u_i f = \langle u_1, \partial_i \rangle u_1 f = 0,$$

whence only the first two terms in the expression of $p_{L_1-L_i}$ act non-trivially. So we get that

$$\mathscr{E}_i f = p_{L_{k-1}-L_k} \cdots p_{L_1-L_{i+1}} \left(1 - \frac{\langle u_i, \partial_1 \rangle \langle u_1, \partial_i \rangle}{\mathbb{E}_1 - \mathbb{E}_i + i - 1 + 1} \right) u_i f.$$

For all $j > i$,

$$\langle u_1, \partial_j \rangle \left(1 - \frac{\langle u_i, \partial_1 \rangle \langle u_1, \partial_i \rangle}{\mathbb{E}_1 - \mathbb{E}_i + i} \right) u_i f = \frac{-1}{\mathbb{E}_1 - \mathbb{E}_i + i - 1} \langle u_1, \partial_j \rangle \langle u_i, \partial_1 \rangle u_1 f$$

$$= \frac{-1}{\mathbb{E}_1 - \mathbb{E}_i + i - 1} (-\langle u_i, \partial_j \rangle + \langle u_i, \partial_1 \rangle \langle u_1, \partial_j \rangle) u_1 f = 0.$$

We can continue the same reasoning, resulting in the embedding factor

$$\mathscr{E}_i = \prod_{j=i-1}^{1} \left(1 - \frac{\langle u_i, \partial_j \rangle \langle u_j, \partial_i \rangle}{\mathbb{E}_j - \mathbb{E}_i + i - j + 1} \right)$$

in its simplest form. Note that the product is ordered from the highest to the lowest index.

6 An Explicit Decomposition of the Twisted Dirac Operator

Now that we know the form of the twistor operators and the embedding factors, we can determine the explicit decomposition of the twisted dirac operator

$$\partial_x^T : \mathscr{C}^\infty(\mathbb{R}^m, \mathscr{S}_\lambda) \to \mathscr{C}^\infty(\mathbb{R}^m, \mathscr{H}_\lambda \otimes \mathbb{S}).$$

This decomposition must be of the form

$$\partial_x^T = \mathcal{Q}_\lambda + \sum_{l=1}^{k} c_l \mathcal{E}_l \mathcal{T}_\lambda^{(l)}.$$

Note that the expressions of the embedding factors do not depend on the space they are acting on, nor on the length of the underlying highest weight. Thus we only have to determine the constants c_l (in terms of Euler operators). In order to gain some insight, we first consider the easy cases. First of all, we of course have the classical Dirac operator

$$\partial_x : \mathscr{C}^\infty(\mathbb{R}^m, \mathbb{S}) \to \mathscr{C}^\infty(\mathbb{R}^m, \mathscr{H}_0 \otimes \mathbb{S}),$$

which is a trivial case, since there are no twistor operators appearing.

6.1 The Case $k = 1$

We have that
$$\partial_x^T : \mathscr{C}^\infty(\mathbb{R}^m, \mathscr{S}_{l_1}) \to \mathscr{C}^\infty(\mathbb{R}^m, \mathscr{H}_{l_1} \otimes \mathbb{S}).$$

One can easily see that the twisted Dirac operator in this case decomposes as follows:

$$\partial_x^T = \left(1 + \frac{u_1 \partial_1}{m + 2\mathbb{E}_1 - 2}\right)\partial_x + \underbrace{\frac{2}{m + 2\mathbb{E}_1 - 2}}_{(a)} \underbrace{u_1}_{(b)} \underbrace{\langle \partial_1, \partial_x \rangle}_{(c)},$$

where we have

(a) the constant c_1,
(b) the embedding factor $\mathcal{E}_1$,
(c) the twistor operator $\mathcal{T}_{l_1}^{(1)}$.

6.2 The Case $k = 2$

In this case, the twisted Dirac operator acts as follows:

$$\partial_x^T : \mathscr{C}^\infty(\mathbb{R}^m, \mathscr{S}_{l_1,l_2}) \to \mathscr{C}^\infty(\mathbb{R}^m, \mathscr{H}_{l_1,l_2} \otimes \mathbb{S}).$$

In order to find the explicit decomposition of this twisted Dirac operator, we need to calculate the constants c_1 and c_2 for which

$$\partial_x^T = \mathcal{Q}_{l_1,l_2} + c_1 \mathcal{E}_1 \mathcal{T}_{l_1,l_2}^{(1)} + c_2 \mathcal{E}_2 \mathcal{T}_{l_1,l_2}^{(2)}.$$

When acting on a function $f \in \mathscr{C}^\infty(\mathbb{R}^m, \mathscr{S}_{l_1,l_2})$, the right-hand side of this equality is given by

$$\left(1 + \frac{u_1 \partial_1}{m + 2\mathbb{E}_1 - 2}\right)\left(1 + \frac{u_2 \partial_2}{m + 2\mathbb{E}_2 - 4}\right)\partial_x f$$

$$+c_1 u_1 \left(1 - \frac{\langle u_2, \partial_1 \rangle \langle u_1, \partial_2 \rangle}{\mathbb{E}_1 - \mathbb{E}_2 + 2}\right)\langle \partial_1, \partial_x \rangle f$$

$$+c_2 \left(1 - \frac{\langle u_2, \partial_1 \rangle \langle u_1, \partial_2 \rangle}{\mathbb{E}_1 - \mathbb{E}_2 + 2}\right)u_2 \langle \partial_2, \partial_x \rangle f.$$

This expression equals

$$\partial_x f - \frac{2}{m + 2\mathbb{E}_1 - 2}u_1\langle \partial_1, \partial_x \rangle f - \frac{2}{m + 2\mathbb{E}_2 - 4}u_2\langle \partial_2, \partial_x \rangle f$$

$$+\frac{4}{(m + 2\mathbb{E}_1 - 2)(m + 2\mathbb{E}_2 - 4)}u_1\langle u_2, \partial_1 \rangle \langle \partial_2, \partial_x \rangle f$$

$$+c_1 u_1\langle \partial_1, \partial_x \rangle f + \frac{c_1}{\mathbb{E}_1 - \mathbb{E}_2 + 1}u_1\langle u_2, \partial_1 \rangle \langle \partial_2, \partial_x \rangle f$$

$$+c_2 u_2\langle \partial_2, \partial_x \rangle f - \frac{c_2}{\mathbb{E}_1 - \mathbb{E}_2 + 2}u_2\langle \partial_2, \partial_x \rangle f$$

$$-\frac{c_2}{\mathbb{E}_1 - \mathbb{E}_2 + 2}u_1\langle u_2, \partial_1 \rangle \langle \partial_2, \partial_x \rangle f.$$

If we combine terms containing the same 'words' (elements of the universal enveloping algebra), we get

$$\partial_x f + \left(c_1 - \frac{2}{m + 2\mathbb{E}_1 - 2}\right)u_1\langle \partial_1, \partial_x \rangle f$$

$$+\left(c_2 - \frac{2}{m + 2\mathbb{E}_2 - 4} - \frac{c_2}{\mathbb{E}_1 - \mathbb{E}_2 + 2}\right)u_2\langle \partial_2, \partial_x \rangle f$$

$$+\left(\frac{4}{(m + 2\mathbb{E}_1 - 2)(m + 2\mathbb{E}_2 - 4)} + \frac{c_1}{\mathbb{E}_1 - \mathbb{E}_2 + 1} - \frac{c_2}{\mathbb{E}_1 - \mathbb{E}_2 + 2}\right)$$

$$\times u_1\langle u_2, \partial_1 \rangle \langle \partial_2, \partial_x \rangle f$$

Here, all coefficients should be zero, except the one of ∂_x. From the second and third coefficient, we get that

$$c_1 = \frac{2}{m + 2\mathbb{E}_1 - 2}$$

and

$$c_2 = \frac{2}{m + 2\mathbb{E}_2 - 4} \frac{\mathbb{E}_1 - \mathbb{E}_2 + 2}{\mathbb{E}_1 - \mathbb{E}_2 + 1}.$$

Substituting these in the final coefficient, it is easily checked that it indeed becomes zero. We thus have found:

$$\partial_x^T = \mathcal{Q}_{l_1,l_2} + \frac{2}{m + 2\mathbb{E}_1 - 2} \mathcal{E}_1 \mathcal{T}_{l_1,l_2}^{(1)} + \frac{2}{m + 2\mathbb{E}_2 - 4} \frac{\mathbb{E}_1 - \mathbb{E}_2 + 2}{\mathbb{E}_1 - \mathbb{E}_2 + 1} \mathcal{E}_2 \mathcal{T}_{l_1,l_2}^{(2)}.$$

6.3 General Case

We now prove the most general case. Consider the twisted Dirac operator

$$\partial_x^T : \mathscr{C}^\infty(\mathbb{R}^m, \mathscr{S}_\lambda) \to \mathscr{C}^\infty(\mathbb{R}^m, \mathscr{H}_\lambda \otimes S),$$

then we want to determine the constants c_l for which

$$\partial_x^T = \mathcal{Q}_\lambda + \sum_{l=1}^{k} c_l \mathcal{E}_l \mathcal{T}_\lambda^{(l)}.$$

To this end, let us first rewrite some operators. For any $f \in \mathscr{C}^\infty(\mathbb{R}^m, \mathscr{S}_\lambda)$, it follows from 3 that

$$\mathcal{Q}_\lambda f = \partial_x f + \sum_{1 \le a_1 < \cdots < a_j \le k} \left(\prod_{i=1}^{j} \frac{-2}{m + 2\mathbb{E}_{a_i} - 2a_i} \right) u_{a_1} \left(\prod_{i=1}^{j-1} \langle u_{a_{i+1}}, \partial_{a_i} \rangle \right) \langle \partial_{a_j}, \partial_x \rangle f.$$

On the other hand, rewriting 4 yields

$$\mathcal{T}_\lambda^{(l)} f = \langle \partial_l, \partial_x \rangle f$$

$$+ \sum_{l = a_1 < \cdots < a_i \le k} \left(\prod_{j=1}^{i} \frac{1}{\mathbb{E}_l - \mathbb{E}_{a_j} + a_j - l + 1} \right) \left(\prod_{j=1}^{i-1} \langle u_{a_{j+1}}, \partial_{a_j} \rangle \right) \langle \partial_{a_i}, \partial_x \rangle f.$$

So we get that

$$\mathscr{E}_l \mathscr{T}^{(l)} = \prod_{b=l-1}^{1} \left(1 - \frac{\langle u_l, \partial_b \rangle \langle u_b, \partial_l \rangle}{\mathbb{E}_b - \mathbb{E}_l + l - b + 1} \right) u_l$$

$$\times \left(\langle \partial_l, \partial_x \rangle + \sum_{l=a_1 < \cdots < a_i \le k} \left(\prod_{j=1}^{i} \frac{1}{\mathbb{E}_l - \mathbb{E}_{a_j} + a_j - l + 1} \right) \right.$$

$$\left. \times \left(\prod_{j=1}^{i-1} \langle u_{a_{j+1}}, \partial_{a_j} \rangle \right) \langle \partial_{a_i}, \partial_x \rangle \right).$$

With these expressions, we can simplify $\mathscr{Q}_\lambda + c_1 \mathscr{E}_1 \mathscr{T}_\lambda^{(1)} + \cdots + c_k \mathscr{E}_k \mathscr{T}_\lambda^{(k)}$. When we use the convention of rewriting each term in this sum in the following form:

(coefficient with Euler operators) $\cdot u_{b_1} \langle u_{b_2}, \partial_{b_1} \rangle \cdots \langle u_{b_i}, \partial_{b_{i-1}} \rangle \langle \partial_{b_i}, \partial_x \rangle$,

with $b_1 < b_2 < \cdots < b_i$, then we get the following coefficients:

- the coefficient of ∂_x equals 1, as we expected;
- the coefficient of $u_1 \langle \partial_1, \partial_x \rangle$ is

$$\frac{-2}{m + 2\mathbb{E}_1 - 2} + c_1,$$

which has to be zero, whence we find $c_1 = \frac{2}{m+2\mathbb{E}_1-2}$;
- similarly, we find that the coefficient of $u_2 \langle \partial_2, \partial_x \rangle$ equals

$$\frac{-2}{m + 2\mathbb{E}_2 - 4} + c_2 - \frac{c_2}{\mathbb{E}_1 - \mathbb{E}_2 + 2 - 1 + 1}.$$

This coefficient also has to equal to zero, whence

$$c_2 = \frac{2}{m + 2\mathbb{E}_2 - 4} \frac{\mathbb{E}_1 - \mathbb{E}_2 + 2}{\mathbb{E}_1 - \mathbb{E}_2 + 1};$$

- for the coefficient of $u_3 \langle \partial_3, \partial_x \rangle$, we find

$$\frac{-2}{m + 2\mathbb{E}_3 - 6} + c_3 - \frac{c_3}{\mathbb{E}_1 - \mathbb{E}_3 + 3} - \frac{c_3}{\mathbb{E}_2 - \mathbb{E}_3 + 2} + \frac{c_3}{(\mathbb{E}_1 - \mathbb{E}_3 + 3)(\mathbb{E}_2 - \mathbb{E}_3 + 2)}$$

$$= \frac{-2}{m + 2\mathbb{E}_3 - 6} + c_3 \left(1 - \frac{1}{\mathbb{E}_1 - \mathbb{E}_3 + 3} \right) \left(1 - \frac{1}{\mathbb{E}_2 - \mathbb{E}_3 + 2} \right),$$

which must equal zero as well, yielding

$$c_3 = \frac{2}{m + 2\mathbb{E}_3 - 6} \frac{\mathbb{E}_1 - \mathbb{E}_3 + 3}{\mathbb{E}_1 - \mathbb{E}_3 + 2} \frac{\mathbb{E}_2 - \mathbb{E}_3 + 2}{\mathbb{E}_2 - \mathbb{E}_3 + 1};$$

- last of all, the general coefficient of $u_l \langle \partial_l, \partial_x \rangle$, for all $l = 1, \ldots, k$ equals

$$
\frac{-2}{m + 2\mathbb{E}_l - 2l} + c_l \left(1 - \frac{1}{\mathbb{E}_1 - \mathbb{E}_l + l - 1 + 1} \right) \cdots
$$

$$
\times \left(1 - \frac{1}{\mathbb{E}_{l-1} - \mathbb{E}_l + l - (l-1) + 1} \right)
$$

or

$$
c_l = \frac{2}{m + 2\mathbb{E}_l - 2l} \prod_{j=1}^{l-1} \left(\frac{\mathbb{E}_j - \mathbb{E}_l + l - j + 1}{\mathbb{E}_j - \mathbb{E}_l + l - j} \right).
$$

This completes the decomposition of the twisted Dirac operator.

7 Conclusion

In this paper, we found the explicit decomposition of the twisted Dirac operator in terms of a higher spin Dirac operator and at most k higher spin twistor operators. To achieve this goal, we needed to calculate the embedding operators $\mathscr{E}_i : \mathscr{S}_{\lambda - L_i} \to \mathscr{H}_\lambda \otimes \mathbb{S}$, as well as a simplicial harmonic projection operator.

References

1. Brackx, F., Delanghe, R., Sommen, F.: Clifford Analysis. Research Notes in Mathematics, vol. 76. Pitman, London (1982)
2. Bureš, J., Sommen, F., Souček, V., Van Lancker, P.: Rarita-Schwinger type operators in Clifford analysis. J. Funct. Anal. **185**, 425–456 (2001)
3. Bureš, J., Sommen, F., Souček, V., Van Lancker, P.: Symmetric analogues of Rarita-Schwinger equations. Ann. Glob. Anal. Geom. **21**(3), 215–240 (2001)
4. Constales, D., Sommen. F., Van Lancker, P.: Models for irreducible representations of Spin(*m*). Adv. Appl. Clifford Algebras **11**(S1), 271–289 (2001)
5. De Schepper, H., Eelbode, D., Raeymaekers, T.: On a special type of solutions for arbitrary higher spin Dirac Operators. J. Phys. A: Math. Theor. **43**(32), 1–13 (2010)
6. De Schepper, H., Eelbode, D., Raeymaekers, T.: Twisted higher spin Dirac operators. Complex Anal. Oper. Theory **8**, 429–447 (2014)
7. Delanghe, R., Sommen, F., Souček, V.: Clifford Analysis and Spinor Valued Functions. Kluwer Academic Publishers, Dordrecht (1992)
8. Dirac, P.A.M.: The quantum theory of the electron. Proc. R. Soc. Lond. **117** (1928)
9. Eelbode, D., Raeymaekers, T.: Construction of higher spin operators using transvector algebras. J. Math. Phys. **55**(10), 101703 (2015)
10. Eelbode, D., Smid, D.: Factorization of Laplace operators on higher spin representations. Complex Anal. Oper. Theory **6**, 1011–1023 (2012)
11. Fegan, H.D.: Conformally invariant first order differential operators. Q. J. Math. **27**, 513–538 (1976)

12. Fulton, W., Harris, J.: Representation Theory: A First Course. Springer, New York (1991)
13. Gilbert, J., Murray, M.A.M.: Clifford Algebras and Dirac Operators in Harmonic Analysis. Cambridge University Press, Cambridge (1991)
14. Howe, R., Tan, E., Willenbring, J.: Stable branching rules for classical symmetric pairs. Trans. AMS **357**(4), 1601–1626 (2004)
15. Humphreys, J.: Introduction to Lie Algebra and Representation Theory. Springer, New York (1972)
16. Klimyk, A.U.: Infinitesimal operators for representations of complex Lie groups and Clebsch-Gordan coefficients for compact groups. J. Phys. A: Math. Gen **15**, 3009–3023 (1982)
17. Molev, A.I.: Yangians and Classical Lie Algebras. Mathematical surveys and monographs, vol. 143. AMS Bookstore (2007)
18. Stein, E.W., Weiss, G.: Generalization of the Cauchy-Riemann equations and representations of the rotation group. Amer. J. Math. **90**, 163–196 (1968)
19. Tolstoy, V.N.: Extremal projections for reductive classical Lie superalgebras with a non-degenerate generalised Killing form. Russ. Math. Surv. **40**, 241–242 (1985)
20. Zhelobenko, D.P.: Transvector algebras in representation theory and dynamic symmetry, group theoretical methods in physics. In: Proceedings of the Third Yurmala Seminar, vol. 1 (1985)

Norms and Moduli on Multicomplex Spaces

M. B. Vajiac

Abstract Multicomplex analysis describes the theory of holomorphic functions on spaces generated by n commuting complex units. In this context we extend some notions of norms and moduli from the space of bicomplex numbers to the space of multicomplex numbers. This approach is meant to be used towards a meaningful theory of Riemannian and semi-Riemannian geometries built on such spaces.

Keywords Bicomplex · Manifolds · Multicomplex analysis · Norms Semi-Riemannian

1 Introduction

The past decade has seen a resurgence of interest in spaces of commuting complex units, their algebra, analysis, especially a theory of holomorphic functions in this sense.

The differential geometry associated to such spaces can be quite useful in modeling quantum mechanics, spaces of harmonic maps and, recently, applications to Teichmuller theory were discovered.

In recent years, some interesting results have been obtained by Baird and Wood [3]. In their work, the theory of harmonic morphisms was linked to a submanifold theory of bicomplex manifolds and used to prove some interesting connections. In this paper, we expand the notion of a bicomplex manifold to the one of a multicomplex one, using the theory of Multicomplex Analysis developed in [22, 24], we revisit some results obtained by Baird and Wood in [3] and expand on them.

In our paper [22] we studied the properties of holomorphic functions of a multicomplex variable and in [24] we developed a hyperfunction theory on the space of holomorphic functions of a multicomplex variable. We studied some of the algebraic properties of the system of differential equations satisfied by such functions, and we

M. B. Vajiac (✉)
Schmid College of Science and Technology, Chapman University, Orange, CA 92866, USA
e-mail: mbvajiac@chapman.edu

© Springer Nature Switzerland AG 2018
P. Cerejeiras et al. (eds.), *Clifford Analysis and Related Topics*,
Springer Proceedings in Mathematics & Statistics 260,
https://doi.org/10.1007/978-3-030-00049-3_7

were able to deduce some new duality theorems which parallel the well known result of Koethe–Martineau–Grothendieck for one and several complex variables.

This paper is written to provide more insight in different types of norms and moduli that can be built on these spaces, with a view towards a differential geometry from this point of view, and reveal related issues and connections that are ripe for discussion. The advantage of these spaces is that they encompass several complex variables into a single variable theory, with possible applications to physics as well as other branches of differential geometry. This paper represents just a step in the discussion of the intricacies of these ambient spaces.

The paper is structured in 8 sections. First we present a theory of bicomplex and multicomplex spaces, then re-iterate a representation theory of bicomplex and multicomplex holomorphic functions.

We then propose different types of norms and moduli for the bicomplex and multicomplex space, extending notions found in our paper [14], where we used the norms on the bicomplex case to extend analyze Berenstein and Lucas type theorems for bicomplex polynomials.

The paper ends with the define the notion of a bicomplex manifold as described by Baird and Wood in [3], followed by the definition and examples of the notion of multicomplex manifolds, together with a structure theorem.

The last section reiterates conclusions and describes a plan of future research into the multiple connections between Multicomplex Analysis and Riemann–Finsler geometry.

2 Bicomplex and Multicomplex Spaces

In this section we will introduce the reader to the algebras of bicomplex and multicomplex spaces. For many more details we refer to the fairly comprehensive recent references [6, 10, 21, 22]. The space $\mathbb{BC}_n$ of multicomplex numbers is the space generated over the reals by n commuting imaginary units. The algebraic properties of this space and analytic properties of multi-complex valued functions defined on $\mathbb{BC}_n$ has been studied in [22, 23]. It is worth noting also that the algebraic properties of the space $\mathbb{BC}_2$ and the properties of its holomorphic functions have been discussed before in [6, 10, 21, 22], using computational algebra techniques. Other related references include [5, 13, 16, 19, 20].

In the case of only one imaginary unit, denoted by $\mathbf{i}_1$, the space $\mathbb{BC}_1$ is the usual complex plane $\mathbb{C}$. Since, in what follows, we will have to work with different complex planes, generated by different imaginary units, we will denote such a space also by $\mathbb{C}(\mathbf{i}_1)$, in order to clarify which is the imaginary unit used in the space itself.

2.1 Bicomplex Space

The next case occurs when we have two commuting imaginary units i_1 and i_2. This yields the bicomplex space $\mathbb{BC}_2$. This space has extensively been studied in e.g. [5, 6, 10, 12, 21], where it is referred to as as $\mathbb{BC}$. We will use this notation throughout the paper whenever we refer to $\mathbb{BC}_2$.

Because of the various units in $\mathbb{BC}_2$, we have several different conjugations that can be defined naturally. Let therefore consider a bicomplex number Z, and let us write it both in terms of its complex coordinates, $Z = z_1 + i_2 z_2$ with $z_1, z_2 \in \mathbb{C}(i_1)$, and in terms of its real coordinates $Z = x_1 + i_1 y_1 + i_2 x_2 + i_1 i_2 y_2$ with $x_\ell, y_\ell \in \mathbb{R}$. Note that $k_{12} := i_1 i_2$ is a hyperbolic unit, i.e. it is a unit which squares to 1. It is worth noting here that from now on all of these spaces will be non-division algebras and we will have a "cone" of zero-divisors described in each case.

A bicomplex number defined as $Z = z_1 + i_2 z_2$ admits several other forms of writing, as follows:

$$
\begin{aligned}
Z &= (x_1 + i_1 y_1) + i_2 (x_2 + i_1 y_2) =: z_1 + i_2 z_2 \\
&= (x_1 + i_2 x_2) + i_1 (y_1 + i_2 y_2) =: \zeta_1 + i_1 \zeta_2 \\
&= (x_1 + k_{12} y_2) + i_1 (y_1 - k_{12} x_2) =: \mathfrak{z}_1 + i_1 \mathfrak{z}_2 \\
&= (x_1 + k_{12} y_2) + i_2 (x_2 - k_{12} y_1) =: \mathfrak{w}_1 + i_2 \mathfrak{w}_2 \\
&= (x_1 + i_1 y_1) + k_{12}(y_2 - i_1 x_2) =: w_1 + k_{12} w_2 \\
&= (x_1 + i_2 x_2) + k_{12}(y_2 - i_2 y_1) =: \omega_1 + k_{12} \omega_2 \\
&= x_1 + i_1 y_1 + i_2 x_2 + k_{12} y_2 \,,
\end{aligned}
$$

where clearly $z_1, z_2, w_1, w_2 \in \mathbb{C}(i_1)$, $\zeta_1, \zeta_2, \omega_1, \omega_2 \in \mathbb{C}(i_2)$, and $\mathfrak{z}_1, \mathfrak{z}_2, \mathfrak{w}_1, \mathfrak{w}_2 \in \mathbb{C}(k_{12})$. Note that we use roman letters such as z, w for complex numbers in $\mathbb{C}(i_1)$, Greek letters ζ, ω for $\mathbb{C}(i_2)$-complex numbers, and Gothic letters $\mathfrak{z}$, $\mathfrak{w}$ for hyperbolic numbers. The hyperbolic number module is denoted usually by $\mathbb{D}(k_{12})$.

We change the traditional notation $\bar{z}$ for complex conjugation in $\mathbb{C}(i_1)$ by $\bar{z}^{i_1}$, corresponding to the involution $i_1 \mapsto -i_1$. Similarly for the conjugation $\bar{\zeta}^{i_2}$ in $\mathbb{C}(i_2)$, corresponding to the involution $i_2 \mapsto -i_2$, and $\bar{\mathfrak{z}}^{k_{12}}$ for the hyperbolic conjugation in $\mathbb{D}(k_{12})$, corresponding to the involution $k_{12} \mapsto -k_{12}$. The traditional bicomplex conjugations are written as follows, using the different writings of bicomplex numbers above:

$$
\begin{aligned}
\overline{Z}^{i_2} &= z_1 - i_2 z_2 = x_1 + i_1 y_1 - i_2 x_2 - k_{12} y_2 \,, \\
\overline{Z}^{i_1} &= \zeta_1 - i_1 \zeta_2 = x_1 - i_1 y_1 + i_2 x_2 - k_{12} y_2 \,, \\
\overline{Z}^{i_1 i_2} &= \mathfrak{z}_1 - i_1 \mathfrak{z}_2 = \mathfrak{w}_1 - i_2 \mathfrak{w}_2 = x_1 - i_1 y_1 - i_2 x_2 + k_{12} y_2 \,.
\end{aligned} \tag{1}
$$

Note that

$$\overline{\left(\overline{Z}^{i_1}\right)}^{i_2} = \overline{Z}^{i_1 i_2},$$

which will be very useful when we will define recursively the multicomplex spaces.

These conjugations yield the following (square of) the moduli of bicomplex numbers:

$$|Z|^2_{i_1} := Z \cdot \overline{Z}^{i_2} = z_1^2 + z_2^2 \in \mathbb{C}(i_1),$$

$$|Z|^2_{i_2} := Z \cdot \overline{Z}^{i_1} = \zeta_1^2 + \zeta_2^2 \in \mathbb{C}(i_2),$$

$$|Z|^2_{k_{12}} := Z \cdot \overline{Z}^{i_1 i_2} = \mathfrak{z}_1^2 + \mathfrak{z}_2^2 \in \mathbb{D}(k_{12}).$$

A function $F : \mathbb{BC}_2 \to \mathbb{BC}_2$ can be written as $F = f_1 + i_2 f_2$, where $f_1, f_2 : \mathbb{BC}_2 \to \mathbb{BC}_1 = \mathbb{C}(i_1)$. The usual complex differentials are defined by

$$\frac{\partial}{\partial z_1} := \frac{1}{2}\left(\frac{\partial}{\partial x_1} - i_1 \frac{\partial}{\partial y_1}\right), \qquad \frac{\partial}{\partial z_2} := \frac{1}{2}\left(\frac{\partial}{\partial x_2} - i_1 \frac{\partial}{\partial y_2}\right),$$

$$\frac{\partial}{\partial \bar{z}_1} := \frac{1}{2}\left(\frac{\partial}{\partial x_1} + i_1 \frac{\partial}{\partial y_1}\right), \qquad \frac{\partial}{\partial \bar{z}_2} := \frac{1}{2}\left(\frac{\partial}{\partial x_2} + i_1 \frac{\partial}{\partial y_2}\right).$$

Following the notations from [22], we introduce the following bicomplex differential operators, written in the standard basis of $\mathbb{BC}$ seen as $\mathbb{C}^2$ in variables z_1 and z_2:

$$\frac{\partial}{\partial Z} := \frac{1}{2}\left(\frac{\partial}{\partial z_1} - i_2 \frac{\partial}{\partial z_2}\right), \qquad \frac{\partial}{\partial \overline{Z}^{i_2}} := \frac{1}{2}\left(\frac{\partial}{\partial z_1} + i_2 \frac{\partial}{\partial z_2}\right),$$

$$\frac{\partial}{\partial \overline{Z}^{i_1}} := \frac{1}{2}\left(\frac{\partial}{\partial \bar{z}_1} - i_2 \frac{\partial}{\partial \bar{z}_2}\right), \qquad \frac{\partial}{\partial \overline{Z}^{i_1 i_2}} := \frac{1}{2}\left(\frac{\partial}{\partial \bar{z}_1} + i_2 \frac{\partial}{\partial \bar{z}_2}\right). \qquad (2)$$

These differential operators have also multiple way of writing, according to the plethora of writings of bicomplex numbers described above.

$\mathbb{BC}_2$ is not a division algebra, and it has two distinguished zero divisors, e_{12} and $\bar{e}_{12}$, which are idempotent, linearly independent over the reals, and mutually annihilating with respect to the bicomplex multiplication:

$$e_{12} := \frac{1 + i_1 i_2}{2}, \qquad \bar{e}_{12} := \frac{1 - i_1 i_2}{2}.$$

Just like $\{1, i_2\}$, they form a basis of the complex algebra $\mathbb{BC}_2$, which is called the *idempotent basis*. Note that the idempotent basis is not an orthogonal basis with respect to the usual Euclidean norm. If we define the following complex variables in $\mathbb{C}(i_1)$:

$$\beta_1 := z_1 - i_1 z_2, \qquad \beta_2 := z_1 + i_1 z_2,$$

the *idempotent representations* for $Z = z_1 + i_2 z_2$ and its conjugates are given by

$$Z = \beta_1 \mathbf{e}_{12} + \beta_2 \bar{\mathbf{e}}_{12}, \qquad \overline{Z}^{i_1} = \bar{\beta}_2 \mathbf{e}_{12} + \bar{\beta}_1 \bar{\mathbf{e}}_{12},$$

$$\overline{Z}^{i_2} = \beta_2 \mathbf{e}_{12} + \beta_1 \bar{\mathbf{e}}_{12}, \qquad \overline{Z}^{i_1 i_2} = \bar{\beta}_1 \mathbf{e}_{12} + \bar{\beta}_2 \bar{\mathbf{e}}_{12}.$$

In the idempotent representation, the bicomplex differential operators defined above become:

$$\frac{\partial}{\partial Z} = \frac{\partial}{\partial \beta_1} \mathbf{e}_{12} + \frac{\partial}{\partial \beta_2} \bar{\mathbf{e}}_{12}, \qquad \frac{\partial}{\partial \overline{Z}^{i_2}} = \frac{\partial}{\partial \beta_2} \mathbf{e}_{12} + \frac{\partial}{\partial \beta_1} \bar{\mathbf{e}}_{12},$$

$$\frac{\partial}{\partial \overline{Z}^{i_1}} = \frac{\partial}{\partial \bar{\beta}_2} \mathbf{e}_{12} + \frac{\partial}{\partial \bar{\beta}_1} \bar{\mathbf{e}}_{12}, \qquad \frac{\partial}{\partial \overline{Z}^{i_1 i_2}} = \frac{\partial}{\partial \bar{\beta}_1} \mathbf{e}_{12} + \frac{\partial}{\partial \bar{\beta}_2} \bar{\mathbf{e}}_{12}.$$

The following notion of a bicomplex derivative is introduced:

Definition 7.1 Let Ω be an open set in $\mathbb{BC}_2$ and let $Z_0 \in \Omega$. A function $F : \Omega \to \mathbb{BC}_2$ is called bicomplex differentiable at Z_0 if the limit

$$\lim_{Z \to Z_0} (Z - Z_0)^{-1} (F(Z) - F(Z_0))$$

exists, for all Z in Ω such that $Z - Z_0$ is invertible, i.e. it is not a divisor of zero. When the limit exists, we will say that the function F has derivative equal to $F'(Z_0) \in \mathbb{BC}_2$ at Z_0.

Note that the limit in the definition above avoids the divisors of zero in $\mathbb{BC}_2$, which are in the union of the two ideals generated by $\mathbf{e}_{12}$ and $\bar{\mathbf{e}}_{12}$, the so-called *cone of singularities*. This set of zero divisors together with 0 is usually denoted by $\mathfrak{S}_0$.

Functions which admit bicomplex derivative at each point in their domain are called bicomplex holomorphic, and it can be shown that this is equivalent to require that they admit a power series expansion in Z [12, Definition 15.2]. However, there are more equivalent statements of bicomplex holomorphy [21]. For example:

Theorem 7.1 *Let Ω be an open set in $\mathbb{BC}_2$ and let $F = f_1 + i_2 f_2 : \Omega \to \mathbb{BC}_2$ be of class $\mathscr{C}^1(\Omega)$. Then F is bicomplex holomorphic on Ω if and only if:*

1. *f_1 and f_2 are complex holomorphic in both complex $\mathbb{C}(i_1)$ variables z_1 and z_2.*
2. *$\dfrac{\partial f_1}{\partial z_1} = \dfrac{\partial f_2}{\partial z_2}$ and $\dfrac{\partial f_2}{\partial z_1} = -\dfrac{\partial f_1}{\partial z_2}$ on Ω; these equations are called the* complex *Cauchy–Riemann conditions.*

Moreover, $F' = \dfrac{1}{2} \dfrac{\partial F}{\partial Z} = \dfrac{\partial f_1}{\partial z_1} + i_2 \dfrac{\partial f_2}{\partial z_1} = \dfrac{\partial f_2}{\partial z_2} - i_2 \dfrac{\partial f_1}{\partial z_2}$, and $F'(Z)$ is invertible if and only if the corresponding Jacobian is non-zero.

As mentioned in [5], the condition $F \in \mathscr{C}^1(U)$ can be dropped via the Hartogs' lemma, on the holomorphicity of functions of several complex variables as being equivalent to their holomorphicity in each variable separately (in other words, no continuity assumption is required). Note here a major difference between quaternionic and bicomplex analysis: as it is well known, in $\mathbb{H}$ the only functions who have quaternionic derivative are quaternionic linear functions, while in the bicomplex setting the class of functions admitting derivative is non-trivial and consists on functions admitting power series expansion.

Similar theorems are valid for the writings $F = \rho_1 + \mathbf{i}\rho_2$, where ρ_1 and ρ_2 are $\mathbb{C}(\mathbf{i}_2)$–valued complex functions of two complex variables (ζ_1, ζ_2) in $\mathbb{C}^2(\mathbf{i}_2)$, and $F = \mathfrak{f}_1 + \mathbf{i}\mathfrak{f}_2$, where $\mathfrak{f}_1$ and $\mathfrak{f}_2$ are $\mathbb{D}(\mathbf{k}_{12})$–valued hyperbolic functions of two hyperbolic variables $(\mathfrak{z}_1, \mathfrak{z}_2)$ in $\mathbb{D}^2(\mathbf{k}_{12})$.

In the idempotent basis, we have the following characterization of bicomplex holomorphy:

Theorem 7.2 *A bicomplex function $F : \Omega \to \mathbb{BC}_2$ is bicomplex holomorphic if and only if*

$$F = G_1(\beta_1)\mathbf{e}_{12} + G_2(\beta_2)\bar{\mathbf{e}}_{12},$$

where G_1 is a complex holomorphic function on the complex domain Ω_1 defined by $\Omega_1 \cdot \mathbf{e}_{12} := \Omega \cdot \mathbf{e}_{12}$, and G_2 is a complex holomorphic function on the complex domain Ω_2 defined by $\Omega_2 \cdot \bar{\mathbf{e}}_{12} := \Omega \cdot \bar{\mathbf{e}}_{12}$.

An immediate consequence of this theorem is that a bicomplex holomorphic function F defined on Ω extends to a bicomplex holomorphic function defined on $\widetilde{\Omega} := \Omega_1 \cdot \mathbf{e}_{12} + \Omega_2 \cdot \bar{\mathbf{e}}_{12}$, which strictly includes Ω. Domains of this form are called bicomplex domains.

A further interesting characterization of holomorphicity in $\mathbb{BC}_2$ is the following result from [6].

Theorem 7.3 *Let $\Omega \subseteq \mathbb{BC}_2$ be an open set and let $F : \Omega \to \mathbb{BC}_2$ be of class $\mathscr{C}^1$ on Ω. Then F is bicomplex holomorphic if and only if F is $\overline{Z}^{\mathbf{i}_1}$, $\overline{Z}^{\mathbf{i}_2}$, $\overline{Z}^{\mathbf{i}_1\mathbf{i}_2}$-regular, i.e.*

$$\frac{\partial F}{\partial \overline{Z}^{\mathbf{i}_1}} = \frac{\partial F}{\partial \overline{Z}^{\mathbf{i}_2}} = \frac{\partial F}{\partial \overline{Z}^{\mathbf{i}_1\mathbf{i}_2}} = 0.$$

Note that both of these last results indicate that holomorphic functions on bicomplex variables can be seen as solutions of overdetermined systems of differential equations with constant coefficients. We have exploited this particularity in our papers [6, 21, 24].

We now turn to the definition of the multicomplex spaces, $\mathbb{BC}_n$, for values of $n \geq 2$. These spaces are defined by taking n commuting imaginary units $\mathbf{i}_1, \mathbf{i}_2, \dots, \mathbf{i}_n$ i.e. $\mathbf{i}_a^2 = -1$, and $\mathbf{i}_a\mathbf{i}_b = \mathbf{i}_b\mathbf{i}_a$ for all a, b. Since the product of two commuting imaginary units is a hyperbolic unit, and since the product of an imaginary unit and a hyperbolic unit is an imaginary unit, we see that these units will generate a space $\mathfrak{A}_n$ of 2^n units, 2^{n-1} of which are imaginary and 2^{n-1} are hyperbolic units. Then the algebra generated over the real numbers by $\mathfrak{A}_n$ is the multicomplex space $\mathbb{BC}_n$ which forms a ring under the usual addition and multiplication operations. As in the $n = 2$ case, the ring $\mathbb{BC}_n$ can be represented as a real algebra, so that each of its elements can be written as $Z = \sum_{I \in \mathfrak{A}_n} Z_I I$, where Z_I are real numbers. This last representation of $Z \in \mathbb{BC}_n$ is quite messy, so we describe a more refined one that will show that these spaces are "nested" spaces.

In particular, following [12], it is natural to define the n-dimensional multicomplex space as follows:

$$\mathbb{BC}_n := \{Z_n = Z_{n-1,1} + \mathbf{i}_n Z_{n-1,2} \mid Z_{n-1,1}, Z_{n-1,2} \in \mathbb{BC}_{n-1}\}$$

with the natural operations of addition and multiplication. Since $\mathbb{BC}_{n-1}$ can be defined in a similar way, we recursively obtain, at the k-th level:

$$Z_n = \sum_{|I|=n-k} \prod_{t=k+1}^{n} (\mathbf{i}_t)^{\alpha_t - 1} Z_{k,I}$$

where $Z_{k,I} \in \mathbb{BC}_k$, $I = (\alpha_{k+1}, \ldots, \alpha_n)$, and $\alpha_j \in \{1, 2\}$.

Because of the existence of n imaginary units, we can define multiple types of conjugations as follows:

$$\overline{Z_n}^{\,\mathbf{i}_l} = \begin{cases} \displaystyle\sum_{|I|=n-k} \prod_{\substack{t=k+1 \\ t \neq l}}^{n} (\mathbf{i}_t)^{\alpha_t - 1} (-\mathbf{i}_l)^{\alpha_l - 1} Z_{k,I} & \text{if } l > k \\[4mm] \displaystyle\sum_{|I|=n-k} \prod_{\substack{t=k+1 \\ t \neq l}}^{n} (\mathbf{i}_t)^{\alpha_t - 1} \overline{Z_{k,I}}^{\,\mathbf{i}_l} & \text{if } l < k \end{cases}$$

and

$$\overline{Z_n}^{\,\mathbf{i}_1 \ldots \mathbf{i}_l} = \overline{\overline{Z_n^{\,\mathbf{i}_1}}^{\,\mathbf{i}_2 \ldots}}^{\,\mathbf{i}_l}$$

It turns out that

$$\overline{Z_n}^{\,\mathbf{i}_l} = \sum_{|I|=n-k} \sum_{i=k+1}^{n} \delta_{l,i}(-1)^{\alpha_l - 1} c_k Z_{k,I} + \sum_{|I|=n-k} \sum_{i=1}^{k} \delta_{l,i} c_k \overline{Z_{k,I}}^{\,\mathbf{i}_l}.$$

Just as in the case of $\mathbb{BC}_2$ we have idempotent bases in $\mathbb{BC}_n$, that will be organized at each "nested" level $\mathbb{BC}_k$ inside $\mathbb{BC}_n$ as follows. Denote by

$$\mathbf{e}_{kl} := \frac{1 + \mathbf{i}_k \mathbf{i}_l}{2}, \qquad \bar{\mathbf{e}}_{kl} := \frac{1 - \mathbf{i}_k \mathbf{i}_l}{2}.$$

Consider the following sets:

$$S_1 := \{\mathbf{e}_{n-1,n}, \bar{\mathbf{e}}_{n-1,n}\},$$
$$S_2 := \{\mathbf{e}_{n-2,n-1} \cdot S_1, \bar{\mathbf{e}}_{n-2,n-1} \cdot S_1\},$$
$$\vdots$$
$$S_{n-1} := \{\mathbf{e}_{12} \cdot S_{n-2}, \bar{\mathbf{e}}_{12} \cdot S_{n-2}\}.$$

At each stage k, the set S_k has 2^k idempotents. Note the following *peculiar* identities:

$$\mathbf{i}_k \cdot \mathbf{e}_{kl} = \frac{\mathbf{i}_k - \mathbf{i}_l}{2} = -\frac{\mathbf{i}_l - \mathbf{i}_k}{2} = -\mathbf{i}_l \cdot \mathbf{e}_{kl},$$
$$\mathbf{i}_k \cdot \bar{\mathbf{e}}_{kl} = \frac{\mathbf{i}_k + \mathbf{i}_l}{2} = \mathbf{i}_l \cdot \bar{\mathbf{e}}_{kl}.$$

It is possible to immediately verify the following

Proposition 7.1 *In each set S_k, the product of any two idempotents is zero.*

We have several idempotent representations of $Z_n \in \mathbb{BC}_n$, as follows.

Theorem 7.4 *Any $Z_n \in \mathbb{BC}_n$ can be written as:*

$$Z_n = \sum_{j=1}^{2^k} Z_{n-k,j} \mathbf{e}_j,$$

where $Z_{n-k,j} \in \mathbb{BC}_{n-k}$ and $\mathbf{e}_j \in S_k$.

Due to the fact that the product of two idempotents is 0 at each level S_k, we will have many zero divisors in $\mathbb{BC}_n$ organized in "singular cones". The topology of the space is difficult, but just like in the case of $\mathbb{BC}_2$ we can circumvent this by avoiding the zero divisors to define the derivative of a multicomplex function as follows.

Definition 7.2 Let Ω be an open set of $\mathbb{BC}_n$ and let $Z_{n,0} \in \Omega$. A function $F : \Omega \to \mathbb{BC}_n$ has a multicomplex derivative at $Z_{n,0}$ if

$$\lim_{Z_n \to Z_{n,0}} (Z_n - Z_{n,0})^{-1} \left(F(Z_n) - F(Z_{n,0}) \right) =: F'(Z_{n,0}),$$

exists whenever $Z_n - Z_{n,0}$ is invertible in $\mathbb{BC}_n$.

Just as in the case of $\mathbb{BC}_2$, functions which admit a multicomplex derivative at each point in their domain are called multicomplex holomorphic, and it can be shown that this is equivalent to require that they admit a power series expansion in Z_n [12, Section 47].

We will denote by $\mathcal{O}(\mathbb{BC}_n)$ the space of multicomplex holomorphic functions. A multicomplex holomorphic function $F \in \mathcal{O}(\Omega)$, where $\Omega \subset \mathbb{BC}_n$, can be split into $F = U + \mathbf{i}_n V$, where U, V are holomorphic functions of two $\mathbb{BC}_{n-1}$ variables. As in the case $n = 2$, there is an equivalent notion of multicomplex holomorphicity, which is more suitable to our computational algebraic purposes, and the following theorem can be proved in a similar fashion as its correspondent for the case $n = 2$ (the differential operators that appear in the statement are defined in detail in the next few sections, but it is not difficult to imagine their actual definition).

Theorem 7.5 *Let Ω be an open set in $\mathbb{BC}_n$ and $F : \Omega \to \mathbb{BC}_n$ such that $F = U + \mathbf{i}_n V \in \mathscr{C}^1(\Omega)$. Then F is multicomplex holomorphic if and only if:*

1. *U and V are multicomplex holomorphic in both multicomplex $\mathbb{BC}_{n-1}$ variables $Z_{n-1,1}$ and $Z_{n-1,2}$.*
2. $\dfrac{\partial U}{\partial Z_{n-1,1}} = \dfrac{\partial V}{\partial Z_{n-1,2}}$ *and* $\dfrac{\partial V}{\partial Z_{n-1,1}} = -\dfrac{\partial U}{\partial Z_{n-1,2}}$ *on Ω; these equations are called the multicomplex Cauchy–Riemann conditions.*

There are several analog statements of the Theorem 7.2 in $\mathbb{BC}_n$, as there are many idempotent representations with respect to each level S_k. For example, for $k = n - 1$, we obtain the following characterization of $\mathbb{BC}_n$–holomorphy:

Theorem 7.6 *A multicomplex function $F : \Omega \to \mathbb{BC}_n$ is $\mathbb{BC}_n$–holomorphic if and only if*

$$F = \sum_{\ell=1}^{2^{n-1}} G_\ell(\beta_\ell)\mathbf{e}_\ell ,$$

where $\mathbf{e}_\ell \in S_{n-1}$ and G_ℓ are complex holomorphic functions on the complex domains Ω_ℓ defined by $\Omega_\ell \cdot \mathbf{e}_\ell := \Omega \cdot \mathbf{e}_\ell$, for all $\ell = 1, \ldots, 2^{n-1}$.

As before, a multicomplex holomorphic function F defined on Ω extends to a multicomplex holomorphic function defined on

$$\widetilde{\Omega} := \sum_{\ell=1}^{2^{n-1}} \Omega_\ell \cdot \mathbf{e}_\ell$$

which strictly includes Ω. Domains of this form are called multicomplex domains.

2.2 The Tricomplex Space $\mathbb{BC}_3$

As an example of the general case, we will explicitly describe holomorphic functions of the space of "tricomplex" numbers $\mathbb{BC}_3$; we will use this space as the first step in the generalization of the bicomplex setting to the multicomplex one.

Let $Z_3 \in \mathbb{BC}_3$ be written as:

$$Z_3 = Z_{21} + i_3 Z_{22} = Z_{111} + i_2 Z_{121} + i_3 Z_{112} + i_2 i_3 Z_{122}$$
$$= x_1 + i_1 x_2 + i_2 x_3 + i_1 i_2 x_4 + i_3 x_5 + i_1 i_3 x_6 + i_2 i_3 x_7 + i_1 i_2 i_3 x_8 .$$

As we indicated earlier, we can define seven different conjugations, depending on which imaginary unit we want to convert to its opposite. We write explicitly a few of them, and leave to the reader to complete the description with the remaining cases:

$$\overline{Z_3}^{i_1} = \overline{Z_{21}}^{i_1} + i_3 \overline{Z_{22}}^{i_1} = \overline{Z_{111}}^{i_1} + i_2 \overline{Z_{121}} + i_3 \overline{Z_{112}} + i_2 i_3 \overline{Z_{122}}$$
$$= x_1 - i_1 x_2 + i_2 x_3 - i_1 i_2 x_4 + i_3 x_5 - i_1 i_3 x_6 + i_2 i_3 x_7 - i_1 i_2 i_3 x_8$$

$$\overline{Z_3}^{i_1 i_2} = \overline{Z_{21}}^{i_1 i_2} + i_3 \overline{Z_{22}}^{i_1 i_2} = \overline{Z_{111}} - i_2 \overline{Z_{121}} + i_3 \overline{Z_{112}} - i_2 i_3 \overline{Z_{122}}$$
$$= x_1 - i_1 x_2 - i_2 x_3 + i_1 i_2 x_4 + i_3 x_5 - i_1 i_3 x_6 - i_2 i_3 x_7 + i_1 i_2 i_3 x_8$$

$$\overline{Z_3}^{i_1 i_2 i_3} = \overline{Z_{21}}^{i_1 i_2} - i_3 \overline{Z_{22}}^{i_1 i_2} = \overline{Z_{111}} - i_2 \overline{Z_{121}} - i_3 \overline{Z_{112}} + i_2 i_3 \overline{Z_{122}}$$
$$= x_1 - i_1 x_2 - i_2 x_3 + i_1 i_2 x_4 - i_3 x_5 + i_1 i_3 x_6 + i_2 i_3 x_7 - i_1 i_2 i_3 x_8$$

In the idempotent representations (there are two of them), consider $Z_3 = Z_{21} + i_3 Z_{22}$, where:

$$Z_{21} = \zeta_{21}^\dagger \mathbf{e}_{12} + \zeta_{21} \bar{\mathbf{e}}_{12}, \qquad Z_{22} = \zeta_{22}^\dagger \mathbf{e}_{12} + \zeta_{22} \bar{\mathbf{e}}_{12} .$$

Then we can write

$$Z_3 = Z_{21} + i_3 Z_{22} = (Z_{21} - i_2 Z_{22})\mathbf{e}_{23} + (Z_{21} + i_2 Z_{22})\bar{\mathbf{e}}_{23}$$
$$= (\zeta_{21}^\dagger + i_1 \zeta_{22}^\dagger)\mathbf{e}_{12}\mathbf{e}_{23} + (\zeta_{21} - i_1 \zeta_{22})\bar{\mathbf{e}}_{12}\mathbf{e}_{23}$$
$$+ (\zeta_{21}^\dagger - i_1 \zeta_{22}^\dagger)\mathbf{e}_{12}\bar{\mathbf{e}}_{23} + (\zeta_{21} + i_1 \zeta_{22})\bar{\mathbf{e}}_{12}\bar{\mathbf{e}}_{23}$$

Therefore, for the conjugates introduced above, we have the following idempotent representations:

$$\overline{Z_3}^{i_1} = \overline{Z_{21}}^{i_1} + i_3 \overline{Z_{22}}^{i_1} = (\overline{Z_{21}}^{i_1} - i_2 \overline{Z_{22}}^{i_1})\mathbf{e}_{23} + (\overline{Z_{21}}^{i_1} + i_2 \overline{Z_{22}}^{i_1})\bar{\mathbf{e}}_{23}$$
$$= (\overline{\zeta_{21}}^{i_1} + i_1 \overline{\zeta_{22}}^{i_1})\mathbf{e}_{12}\mathbf{e}_{23} + (\overline{\zeta_{21}^\dagger}^{i_1} - i_1 \overline{\zeta_{22}^\dagger}^{i_1})\bar{\mathbf{e}}_{12}\mathbf{e}_{23}$$
$$+ (\overline{\zeta_{21}}^{i_1} - i_1 \overline{\zeta_{22}}^{i_1})\mathbf{e}_{12}\bar{\mathbf{e}}_{23} + (\overline{\zeta_{21}^\dagger}^{i_1} + i_1 \overline{\zeta^\dagger_{22}}^{i_1})\bar{\mathbf{e}}_{12}\bar{\mathbf{e}}_{23}$$

$$\overline{Z_3}^{i_1 i_2} = \overline{Z_{21}}^{i_1 i_2} + i_3 \overline{Z_{22}}^{i_1 i_2} = (\overline{Z_{21}}^{i_1 i_2} - i_2 \overline{Z_{22}}^{i_1 i_2})\mathbf{e}_{23} + (\overline{Z_{21}}^{i_1 i_2} + i_2 \overline{Z_{22}}^{i_1 i_2})\bar{\mathbf{e}}_{23}$$
$$= (\overline{\zeta_{21}^\dagger}^{i_1} + i_1 \overline{\zeta_{22}^\dagger}^{i_1})\mathbf{e}_{12}\mathbf{e}_{23} + (\overline{\zeta_{21}}^{i_1} - i_1 \overline{\zeta_{22}}^{i_1})\bar{\mathbf{e}}_{12}\mathbf{e}_{23}$$
$$+ (\overline{\zeta_{21}^\dagger}^{i_1} - i_1 \overline{\zeta_{22}^\dagger}^{i_1})\mathbf{e}_{12}\bar{\mathbf{e}}_{23} + (\overline{\zeta_{21}}^{i_1} + i_1 \overline{\zeta_{22}}^{i_1})\bar{\mathbf{e}}_{12}\bar{\mathbf{e}}_{23}$$

$$\overline{Z}_3^{i_1 i_2 i_3} = \overline{Z}_{21}^{i_1 i_2} - \mathbf{i}_3 \overline{Z}_{22}^{i_1 i_2} = (\overline{Z}_{21}^{i_1 i_2} + \mathbf{i}_2 \overline{Z}_{22}^{i_1 i_2})\mathbf{e}_{23} + (\overline{Z}_{21}^{i_1 i_2} - \mathbf{i}_2 \overline{Z}_{22}^{i_1 i_2})\mathbf{e}_{\overline{23}}$$

$$= (\zeta_{21}^{\dagger i_1} - \mathbf{i}_1 \zeta_{22}^{\dagger i_1})\mathbf{e}_{12}\mathbf{e}_{23} + (\zeta_{21}^{- i_1} + \mathbf{i}_1 \zeta_{22}^{- i_1})\mathbf{e}_{\overline{12}}\mathbf{e}_{23}$$

$$+ (\zeta_{21}^{\dagger i_1} + \mathbf{i}_1 \zeta_{22}^{\dagger i_1})\mathbf{e}_{12}\mathbf{e}_{\overline{23}} + (\zeta_{21}^{- i_1} - \mathbf{i}_1 \zeta_{22}^{- i_1})\mathbf{e}_{\overline{12}}\mathbf{e}_{\overline{23}},$$

and similarly for the other four conjugates $\overline{Z}_3^{i_2}, \overline{Z}_3^{i_3}, \overline{Z}_3^{i_1 i_3}, \overline{Z}_3^{i_2 i_3}$. The tricomplex differential operators are defined as follows:

$$\frac{\partial}{\partial Z_3}^{(3)} = \frac{\partial}{\partial Z_{21}}^{(2)} - \mathbf{i}_3 \frac{\partial}{\partial Z_{22}}^{(2)} = \frac{\partial}{\partial Z_{111}} - \mathbf{i}_2 \frac{\partial}{\partial Z_{121}} - \mathbf{i}_3 \frac{\partial}{\partial Z_{112}} + \mathbf{i}_2 \mathbf{i}_3 \frac{\partial}{\partial Z_{122}}$$

$$\frac{\partial}{\partial \overline{Z}_3^{i_1}}^{(3)} = \frac{\partial}{\partial \overline{Z}_{21}^{i_1}}^{(2)} - \mathbf{i}_3 \frac{\partial}{\partial \overline{Z}_{22}^{i_1}}^{(2)} = \frac{\partial}{\partial \overline{Z}_{111}} - \mathbf{i}_2 \frac{\partial}{\partial \overline{Z}_{121}} - \mathbf{i}_3 \frac{\partial}{\partial \overline{Z}_{112}} + \mathbf{i}_2 \mathbf{i}_3 \frac{\partial}{\partial \overline{Z}_{122}}$$

$$\frac{\partial}{\partial \overline{Z}_3^{i_1 i_2}}^{(3)} = \frac{\partial}{\partial \overline{Z}_{21}^{i_1 i_2}}^{(2)} - \mathbf{i}_3 \frac{\partial}{\partial \overline{Z}_{22}^{i_1 i_2}}^{(2)} = \frac{\partial}{\partial \overline{Z}_{111}} + \mathbf{i}_2 \frac{\partial}{\partial \overline{Z}_{121}} - \mathbf{i}_3 \frac{\partial}{\partial \overline{Z}_{112}} - \mathbf{i}_2 \mathbf{i}_3 \frac{\partial}{\partial \overline{Z}_{122}}$$

$$\frac{\partial}{\partial \overline{Z}_3^{i_1 i_2 i_3}}^{(3)} = \frac{\partial}{\partial \overline{Z}_{21}^{i_1 i_2}}^{(2)} + \mathbf{i}_3 \frac{\partial}{\partial \overline{Z}_{22}^{i_1 i_2}}^{(2)} = \frac{\partial}{\partial \overline{Z}_{111}} + \mathbf{i}_2 \frac{\partial}{\partial \overline{Z}_{121}} + \mathbf{i}_3 \frac{\partial}{\partial \overline{Z}_{112}} + \mathbf{i}_2 \mathbf{i}_3 \frac{\partial}{\partial \overline{Z}_{122}}$$

where the upper indices are placeholders for the multicomplex space the operators belong to, with similar forms for the other four differential operators

$$\frac{\partial}{\partial \overline{Z}_3^{i_2}}^{(3)}, \quad \frac{\partial}{\partial \overline{Z}_3^{i_3}}^{(3)}, \quad \frac{\partial}{\partial \overline{Z}_3^{i_2 i_3}}^{(3)}, \quad \frac{\partial}{\partial \overline{Z}_3^{i_1 i_3}}^{(3)}.$$

Let now $F : \Omega \subset \mathbb{BC}_3 \to \mathbb{BC}_3$ be a function that we write as usual as

$$F = U + \mathbf{i}_3 V = u_1 + \mathbf{i}_2 v_1 + \mathbf{i}_3 u_2 + \mathbf{i}_2 \mathbf{i}_3 v_2$$

$$= f_1 + \mathbf{i}_1 f_2 + \mathbf{i}_2 f_3 + \mathbf{i}_1 \mathbf{i}_2 f_4 + \mathbf{i}_3 f_5 + \mathbf{i}_1 \mathbf{i}_3 f_6 + \mathbf{i}_2 \mathbf{i}_3 f_7 + \mathbf{i}_1 \mathbf{i}_2 \mathbf{i}_3 f_8$$

For simplicity of notation we will write Z instead of Z_{21} and W instead of Z_{22}. Now we study the system formed by all seven differentials of F equal 0. The equation $\frac{\partial}{\partial \overline{Z}_3^{i_1}}^{(3)} F = 0$ is equivalent to:

$$\left(\frac{\partial}{\partial \overline{Z}^{i_1}} - \mathbf{i}_3 \frac{\partial}{\partial \overline{W}^{i_1}} \right) (U + \mathbf{i}_3 V) = 0$$

which is equivalent to the Cauchy–Riemann type system

$$\frac{\partial}{\partial \overline{Z}^{i_1}} U + \frac{\partial}{\partial \overline{W}^{i_1}} V = 0, \qquad \frac{\partial}{\partial \overline{W}^{i_1}} U - \frac{\partial}{\partial \overline{Z}^{i_1}} V = 0.$$

The equation $\dfrac{\partial}{\partial \bar{Z}_3^{\mathbf{i}_2}}^{(3)} F = 0$ is equivalent to

$$\left(\frac{\partial}{\partial \bar{Z}^{\mathbf{i}_2}} - \mathbf{i}_3 \frac{\partial}{\partial \bar{W}^{\mathbf{i}_2}} \right) (U + \mathbf{i}_3 V) = 0,$$

which is equivalent to the Cauchy–Riemann type system:

$$\frac{\partial}{\partial \bar{Z}^{\mathbf{i}_2}} U + \frac{\partial}{\partial \bar{W}^{\mathbf{i}_2}} V = 0, \qquad \frac{\partial}{\partial \bar{W}^{\mathbf{i}_2}} U - \frac{\partial}{\partial \bar{Z}^{\mathbf{i}_2}} V = 0.$$

One can then argue in exactly the same way for the remaining five differential operators, and we obtain the following important result.

Theorem 7.7 *A function $F = U + \mathbf{i}_3 V : \Omega \subset BC_3 \to \mathbb{BC}_3$ is in the kernel of all 7 differential operators generated by the seven conjugations, if and only if U and V are $\mathbb{BC}_2$-holomorphic functions satisfying the bicomplex Cauchy–Riemann conditions.*

Proof We refer the reader to [22, 24] for this proof.

3 Representations of Bicomplex Holomorphic Functions

Following [23], in the theory of bicomplex functions, the complex light cone in two dimensions, i.e. $\Gamma = \{(z_1, z_2) \mid z_1^2 + z_2^2 = 0\}$, plays a very important role as it coincides with the set $\mathfrak{S}_0$ of zero-divisors in $\mathbb{BC}_2$ (together with 0). The complex Laplacian plays a prominent role because it can be factored as the product of two linear operators, one of which is the bicomplex differentiation.

The following are candidates for being the Laplacians in $\mathbb{BC}_2$:

$$\Delta_{\mathbb{C}^2(\mathbf{i}_1)} := \frac{\partial^2}{\partial z_1^2} + \frac{\partial^2}{\partial z_2^2}; \tag{3}$$

$$\Delta_{\mathbb{C}^2(\mathbf{i}_2)} := \frac{\partial^2}{\partial \zeta_1^2} + \frac{\partial^2}{\partial \zeta_2^2}; \tag{4}$$

$$\Delta_{\mathbb{D}(\mathbf{k}_{12})} := \frac{\partial^2}{\partial \mathfrak{z}_1^2} + \frac{\partial^2}{\partial \mathfrak{z}_2^2}. \tag{5}$$

Operators (3) and (4) are called complex ($\mathbb{C}(\mathbf{i}_1)$ and $\mathbb{C}(\mathbf{i}_2)$ respectively) Laplacians and (5) is the hyperbolic Laplacian. The first of them acts on $\mathbb{C}(\mathbf{i}_1)$-valued holomorphic functions of two complex variables z_1 and z_2; the second acts on $\mathbb{C}(\mathbf{i}_2)$-valued holomorphic functions of the complex variables ζ_1 and ζ_2; and the third acts on $\mathbb{D}(\mathbf{k}_{12})$-valued holomorphic functions of the hyperbolic variables $\mathfrak{z}_1$ and $\mathfrak{z}_2$.

It turns out that

$$\Delta_{\mathbb{C}^2(\mathbf{i}_1)} = 4 \frac{\partial}{\partial Z} \frac{\partial}{\partial Z^\dagger}, \tag{6}$$

where the operators act on $\mathbb{BC}_2$-valued functions holomorphic in the sense of the complex variables z_1, z_2. Hence, the theory of $\mathbb{BC}_2$−holomorphic functions can be seen as the function theory for $\mathbb{C}(\mathbf{i}_1)$-complex Laplacian. This factorization allows us to establish direct relations between $\mathbb{BC}_2$−holomorphic functions and complex harmonic functions, that is, null solutions to the operator $\Delta_{\mathbb{C}^2(\mathbf{i})}$.

As we mentioned before, bicomplex holomorphic functions are naturally connected to a certain Cauchy–Riemann type system. To be more precise, a function $F : \Omega \subset \mathbb{BC}_2 \to \mathbb{BC}_2$ can be written as

$$F(z_1, z_2) = f_1(z_1, z_2) + \mathbf{i}_2 f_2(z_1, z_2),$$

where the identification between $\mathbb{BC}_2$ and $\mathbb{C}^2(\mathbf{i}_1)$ is given by $Z = z_1 + \mathbf{i}_2 z_2 \mapsto (z_1, z_2)$. Recall that the condition of bicomplex holomorphicity on F is equivalent to the request that f_1 and f_2 are holomorphic functions of two complex variables, satisfying the additional Cauchy–Riemann type system

$$\frac{\partial f_1}{\partial z_1} = \frac{\partial f_2}{\partial z_2} \quad \text{and} \quad \frac{\partial f_2}{\partial z_1} = -\frac{\partial f_1}{\partial z_2}. \tag{7}$$

Denote by $\Omega \to \mathscr{B}_2(\Omega)$ the sheaf of bicomplex holomorphic functions; we then have the following isomorphisms of function spaces:

$$\mathscr{B}_2(\mathbb{BC}) \simeq \left[\mathscr{O}(\mathbb{C}^2) \times \mathscr{O}(\mathbb{C}^2) \right]^{\partial_{z^{\mathbf{i}_2}}}$$

where recall that

$$\partial_{Z^{\mathbf{i}_2}} := \frac{\partial}{\partial Z^{\mathbf{i}_2}} = \frac{\partial}{\partial z_1} + \mathbf{i}_2 \frac{\partial}{\partial z_2}$$

is the main conjugate bicomplex differential operator acting on bicomplex functions $F = f_1 + \mathbf{i}_2 f_2$ by

$$f_1 + \mathbf{i}_2 f_2 \mapsto (\partial_{z_1} f_1 - \partial_{z_2} f_2, \partial_{z_2} f_1 + \partial_{z_1} f_2).$$

Rewriting the functions as:

$$f_1(z_1, z_2) = g_1(z_1 - \mathbf{i}_1 z_2) + g_2(z_1 + \mathbf{i}_1 z_2), \quad f_2(z_1, z_2) = h_1(z_1 - \mathbf{i}_1 z_2) + h_2(z_1 + \mathbf{i}_1 z_2),$$

and redefining the variables as: $\beta_1 = z_1 - i_1 z_2$ and $\beta_2 = z_1 + i_1 z_2$, then the first Cauchy–Riemann equation in (7) is enough to imply

$$h_1(\beta_1) = i_1 g_1(\beta_1), \qquad h_2(\beta_2) = -i_1 g_2(\beta_2),$$

up to a complex constant.

In the idempotent representation of bicomplex numbers and functions: we write $Z = \beta_1 \mathbf{e}_{12} + \beta_2 \bar{\mathbf{e}}_{12}$ and $F = u_1 \mathbf{e}_{12} + u_2 \bar{\mathbf{e}}_{12}$, where

$$\beta_1 = z_1 - i_1 z_2 \quad \text{and} \quad \beta_2 = z_1 + i_1 z_2$$

are $\mathbb{C}(i_1)$–complex numbers, and

$$u_1 := f_1 - i_1 f_2 \quad \text{and} \quad u_2 := f_1 + i_1 f_2$$

are complex functions in β_1 and β_2. Then the operator $\partial_{\bar{Z}^{12}}$ is given by

$$\partial_{\bar{Z}^{12}} = \partial_{\beta_2} \mathbf{e}_{12} + \partial_{\beta_1} \bar{\mathbf{e}}_{12}$$

acting on $F = (u_1, u_2)$.

Note that in the idempotent coordinates, we have:

$$\mathscr{B}_2(\mathbb{BC}) \simeq \left[\mathscr{O}(\mathbb{C}^2) \times \mathscr{O}(\mathbb{C}^2) \right]^{\partial_{\beta_2} \mathbf{e}_{12} + \partial_{\beta_1} \bar{\mathbf{e}}_{12}}$$

In a similar way, for $\mathscr{B}_n(\mathbb{BC}_n)$ the sheaf of multicomplex holomorphic functions on the multicomplex space we can write:

$$\mathscr{B}_n(\mathbb{BC}_n) \simeq \left[\mathscr{O}(\mathbb{C}^2) \times \cdots \times \mathscr{O}(\mathbb{C}^2) \right]^{\partial_{\beta_1} \mathbf{e}_1 + \cdots + \partial_{\beta_{2^n-1}} \mathbf{e}_{2^n-1}}$$

For details we invite the reader to consult [23].

4   Norms and Moduli on the Bicomplex Space

We would like to point out some results proven in [14] that indicate that the Euclidean norm is not the right object in analyzing a bicomplex space. We start with introducing the set of hyperbolic numbers and re-iterating the norms found in [14], where we elaborate on the several types of norms one can define on the space of bicomplex numbers. Let $|z| = \|(x, y)\|$ be the usual Euclidean norm on $\mathbb{C}_{i_1} \simeq \mathbb{R}^2$. The open and closed disks in $\mathbb{C}_{i_1}$ (or $\mathbb{C}_{i_2}$) are denoted by $B_z(r)$ and $\bar{B}_z(r)$, for $0 < r \leq \infty$.

4.1 The Euclidean Norm

The Euclidean norm $\|Z\|$ on $\mathbb{BC}_2$, when it is seen as $\mathbb{C}^2(\mathbf{i}_1)$, $\mathbb{C}^2(\mathbf{i}_2)$ or $\mathbb{R}^4$ is:

$$\|Z\| = \sqrt{|z_1|^2 + |z_2|^2} = \sqrt{Re\left(|Z|^2_{\mathbf{i}_1\mathbf{i}_2}\right)} = \sqrt{x_1^2 + y_1^2 + x_2^2 + y_2^2}.$$

As studied in detail in [10], in idempotent coordinates $Z = \beta_1 \mathbf{e}_{12} + \beta_2 \bar{\mathbf{e}}_{12}$, the Euclidean norm becomes:

$$\|Z\| = \frac{1}{\sqrt{2}}\sqrt{|\beta_1|^2 + |\beta_2|^2}. \tag{8}$$

The Euclidean ball of radius 1 in the bicomplex space is a subset of the cartesian product of the complex balls of radius $\sqrt{2}$ in β_1 and β_2, respectively. Albeit this is the easiest to define, the Euclidean norm fails to capture the complexities of the space. It is easy to prove that

$$\|Z \cdot W\| \le \sqrt{2}\left(\|Z\| \cdot \|W\|\right), \tag{9}$$

and we note that this inequality is sharp since if $Z = W = \mathbf{e}$, one has:

$$\|\mathbf{e}_{12} \cdot \mathbf{e}_{12}\| = \|\mathbf{e}_{12}\| = \frac{1}{\sqrt{2}} = \sqrt{2}\,\|\mathbf{e}_{12}\| \cdot \|\mathbf{e}_{12}\|.$$

We also have the following results about the multiplication property of the Euclidean norm on bicomplex numbers (see [10] for details):

Proposition 7.2 *If $Z = \beta_1 \mathbf{e}_{12} + \beta_2 \bar{\mathbf{e}}_{12}$ and $W = \gamma_1 \mathbf{e}_{12} + \gamma_2 \bar{\mathbf{e}}_{12}$ are two bicomplex numbers then $\|Z \cdot W\| = \|Z\| \cdot \|W\|$ if and only if*

$$|\beta_1| = |\beta_2| \quad or \quad |\gamma_1| = |\gamma_2|.$$

Proposition 7.3 *The Euclidean norm of the product of two bicomplex numbers is equal to the product of their norms if and only if at least one of them is the product of a complex number in $\mathbb{C}_{\mathbf{i}_1}$ and of a complex number in $\mathbb{C}_{\mathbf{i}_2}$.*

4.2 The Set of Hyperbolic Numbers and the Hyperbolic Valued Norm

A special subalgebra of $\mathbb{BC}_2$ is the set of hyperbolic numbers. The algebra and the analysis of hyperbolic numbers have been studied, for example, in e.g. [1, 4, 10], we summarize below only the notions relevant for our results. A hyperbolic number can be defined independently of $\mathbb{BC}$, by $\mathfrak{z} = x + \mathbf{k}y$, with $x, y, \in \mathbb{R}, \mathbf{k} \notin \mathbb{R}, \mathbf{k}^2 = 1$, and we denote by d the algebra of hyperbolic numbers with the usual component–wise addition and multiplication. The hyperbolic *conjugate* of $\mathfrak{z}$ is defined by $\mathfrak{z}^\circ :=$ $x - \mathbf{k}y$, and note that:

$$\mathfrak{z} \cdot \mathfrak{z}^\circ = x^2 - y^2 \in \mathbb{R}, \tag{10}$$

which yields the notion of the square of the *modulus* of a hyperbolic number $\mathfrak{z}$, defined by $|\mathfrak{z}|_{\mathsf{d}}^2 := \mathfrak{z} \cdot \mathfrak{z}^\circ$. Similar to the bicomplex case, hyperbolic numbers have a unique idempotent representation with real coefficients, which, when d is viewed as embedded in $\mathbb{BC}_2$, namely $\mathbf{k} = \mathbf{i}_1 \mathbf{i}_2$ we have:

$$\mathfrak{z} = s\mathbf{e}_{12} + t\bar{\mathbf{e}}_{12}. \tag{11}$$

where $\mathbf{e}_{12} = \dfrac{1}{2}(1 + \mathbf{k}) = \dfrac{1}{2}(1 + \mathbf{i}_1 \mathbf{i}_2)$, $\bar{\mathbf{e}}_{12} = \dfrac{1}{2}(1 - \mathbf{k}) = \dfrac{1}{2}(1 - \mathbf{i}_1 \mathbf{i}_2)$, and where $s := x + y$ and $t := x - y$. Note that $\mathbf{e}_{12}^\circ = \bar{\mathbf{e}}_{12}$ if we consider d as a subset of $\mathbb{BC}_2$, as briefly explained above. Observe that

$$|\mathfrak{z}|_{\mathsf{d}}^2 = x^2 - y^2 = (x + y)(x - y) = st.$$

Define the set d^+ of non-negative hyperbolic numbers by:

$$\mathsf{d}^+ = \left\{ x + \mathbf{k}y \,|\, x^2 - y^2 \geq 0, x \geq 0 \right\} = \left\{ x + \mathbf{k}y \,|\, x \geq 0, |y| \leq x \right\}$$
$$= \{ s\mathbf{e}_{12} + t\bar{\mathbf{e}}_{12} \,|\, s, t \geq 0 \}.$$

The hyperbolic algebra d is a subalgebra of the bicomplex numbers $\mathbb{BC}_2$ (see [10] for details). Actually $\mathbb{BC}_2$ is the algebraic closure of d, and it can also be seen as the complexification of d by using either of the imaginary unit $\mathbf{i}_1$ or the unit $\mathbf{i}_2$.

As studied extensively in [1], one can define a partial order relation defined on d by:

$$\mathfrak{z}_1 \preceq \mathfrak{z}_2 \quad \text{if and only if} \quad \mathfrak{z}_2 - \mathfrak{z}_1 \in \mathsf{d}^+. \tag{12}$$

In the following section we will use this partial order to study the *hyperbolic-valued* norm, which was first introduced and studied in [1].

Using the partial order relation (12) on d, one can define a *hyperbolic-valued* norm for $\mathfrak{z} = x + \mathbf{k}y = s\mathbf{e}_{12} + t\bar{\mathbf{e}}_{12}$ by:

$$|\mathfrak{z}|_h := |s|\mathbf{e}_{12} + |t|\bar{\mathbf{e}}_{12} \in \mathsf{d}^+.$$

It is shown in [1] that this definition obeys the corresponding properties of a norm, i.e. $|\mathfrak{z}|_h = 0$ if and only if $\mathfrak{z} = 0$, it is multiplicative: $|\mathfrak{z}_1 \mathfrak{z}_2|_h = |\mathfrak{z}_1|_h \cdot |\mathfrak{z}_2|_h$, for any real number $\lambda \in \mathbb{R}$, we have $|\lambda \mathfrak{z}|_h = |\lambda| \cdot |\mathfrak{z}|_h$, and it respects the triangle inequality with respect to the order introduced above:

$$|\mathfrak{z}_1 + \mathfrak{z}_2|_h \preceq |\mathfrak{z}_1|_h + |\mathfrak{z}_2|_h,$$

for all $\mathfrak{z}_1, \mathfrak{z}_2 \in d$. The hyperbolic-valued norm coincides with the modulus $|Z|_\mathbf{k}$ of a hyperbolic number $Z \in d \subset \mathbb{BC}_2$. Indeed, recall that the hyperbolic-valued modulus $|Z|_\mathbf{k}$ of a bicomplex number $Z = \beta_1 \mathbf{e}_{12} + \beta_2 \bar{\mathbf{e}}_{12}$, given by

$$|Z|_\mathbf{k}^2 = |\zeta_1|^2 \mathbf{e}_{12} + |\zeta_2|^2 \bar{\mathbf{e}}_{12},$$

is a non-negative hyperbolic number in d^+. Moreover, one can properly define the square root of a hyperbolic number (see [1, 10]), hence the modulus $|Z|_\mathbf{k}$ is well-defined and given by:

$$|Z|_\mathbf{k} := |\zeta_1| \mathbf{e}_{12} + |\zeta_2| \bar{\mathbf{e}}_{12} \in d^+.$$

In the case $Z \in d \subset \mathbb{BC}_2$, i.e. $\beta_1, \beta_2 \in \mathbb{R}$, it follows that $|Z|_h = |Z|_{\mathbf{i}_1 \mathbf{i}_2}$.

4.3 Finsler-Type Modulus

Another real-valued fourth–degree modulus that it is used in this setting is the one found by multiplying a bicomplex number by all its conjugates:

$$|Z|_\mathscr{F}^4 := Z\, \overline{Z}^{\mathbf{i}_1}\, \overline{Z}^{\mathbf{i}_2}\, \overline{Z}^{\mathbf{i}_1 \mathbf{i}_2}.$$

This is a very useful modulus that is related to the fourth order Laplacian in the bicomplex case. One can easily check that $|Z|_\mathscr{F}^4$ is indeed a real number, it is positive when Z is not a divisor of 0, and it is 0 when $Z \in \Sigma_0$, i.e. Z is a complex multiple of $\mathbf{e}_{12}$ or $\bar{\mathbf{e}}_{12}$. These results are best seen in the idempotent representation of bicomplex numbers. Indeed, recall that if $Z = \beta_1 \mathbf{e}_{12} + \zeta_2 \bar{\mathbf{e}}_{12}$, with $\beta_1, \beta_2 \in \mathbb{C}_{\mathbf{i}_1}$, i.e.:

$$\beta_1 := z_1 - \mathbf{i}_1 z_2, \qquad \beta_2 := z_1 + \mathbf{i}_1 z_2,$$

then the *idempotent representations* for $Z = z_1 + \mathbf{i}_2 z_2$ and its conjugates are given by

$$Z = \beta_1 \mathbf{e}_{12} + \beta_2 \bar{\mathbf{e}}_{12}, \qquad \overline{Z}^{\mathbf{i}_1} = \overline{\beta}_2^{\mathbf{i}_1} \mathbf{e}_{12} + \overline{\beta}_1^{\mathbf{i}_1} \bar{\mathbf{e}}_{12},$$
$$\overline{Z}^{\mathbf{i}_2} = \beta_2 \mathbf{e}_{12} + \beta_1 \bar{\mathbf{e}}_{12}, \qquad \overline{Z}^{\mathbf{i}_1 \mathbf{i}_2} = \overline{\beta}_1^{\mathbf{i}_1} \mathbf{e}_{12} + \overline{\beta}_2^{\mathbf{i}_1} \bar{\mathbf{e}}_{12}.$$

This leads to

$$|Z|_\mathscr{F}^4 = |\beta_1|^2 \cdot |\beta_2|^2 \in \mathbb{R}^+.$$

Equivalently:

$$|Z|_\mathscr{F} = \sqrt{|\beta_1| \cdot |\beta_2|} \in \mathbb{R}^+. \tag{13}$$

Note that $|Z|_{\mathscr{F}} = 0$ if and only if $\beta_1 = 0$ or $\beta_2 = 0$, i.e. Z is a zero divisor, or both are equal zero, i.e. $Z = 0$. Moreover, if $\lambda \in \mathbb{C}_{i_1}$ then

$$|\lambda Z|_{\mathscr{F}} = \sqrt{|\lambda \beta_1| \cdot |\lambda \beta_2|} = |\lambda|\sqrt{|\beta_1| \cdot |\beta_2|} = |\lambda||Z|_{\mathscr{F}}.$$

Consider $Z = \beta_1 \mathbf{e}_{12} + \beta_2 \bar{\mathbf{e}}_{12}$ and $W = \eta_1 \mathbf{e}_{12} + \eta_2 \bar{\mathbf{e}}_{12}$. Then

$$|Z + W|_{\mathscr{F}} = \sqrt{|\beta_1 + \eta_1| \cdot |\beta_2 + \eta_2|}.$$

This modulus is less tractable than the other norms defined in this section. For example, the author has found that Bernstein and Lucas type inequalities for bicomplex polynomials do not hold in this case and restrictions on which they may hold are studied now.

However, if one wants to define a theory of harmonic maps for the fourth order Laplacian in the bicomplex case, this modulus fits the problem best and it is worth further study.

4.4   Lie and Dual Lie Norms

The Lie and dual Lie norms are usually defined and studied in the context of several complex variables, see for example [11, 18]. A bicomplex number $Z = z_1 + i_2 z_2$ can be viewed as a two–complex vector, therefore the cross norm $L(Z)$ on $\mathbb{BC}_2$ corresponding to $Z = (z_1, z_2) \in \mathbb{C}_{i_1}^2$ is the Lie norm given by:

$$L(Z) := \sqrt{x_1^2 + x_2^2 + y_1^2 + y_2^2 + 2\sqrt{(x_1^2 + x_2^2)(y_1^2 + y_2^2) - (x_1 y_1 + x_2 y_2)^2}}$$

$$= \sqrt{x_1^2 + x_2^2 + y_1^2 + y_2^2 + 2|x_1 y_2 - x_2 y_1|}. \tag{14}$$

Consider now Z written in the form:

$$Z = (x_1 + i_1 i_2 y_2) + i_1(y_1 - i_1 i_2 x_2) =: \mathfrak{z}_1 + i_1 \mathfrak{z}_2,$$

and write $\mathfrak{z}_1$ and $\mathfrak{z}_2$ in their hyperbolic idempotent representations:

$$\mathfrak{z}_1 = (x_1 + y_2)\mathbf{e}_{12} + (x_1 - y_2)\bar{\mathbf{e}}_{12} =: s_1 \mathbf{e}_{12} + t_1 \bar{\mathbf{e}}_{12},$$
$$\mathfrak{z}_2 = (y_1 - x_2)\mathbf{e}_{12} + (y_1 + x_2)\bar{\mathbf{e}}_{12} =: s_2 \mathbf{e}_{12} + t_2 \bar{\mathbf{e}}_{12}.$$

In (14), if $x_1 y_2 - x_2 y_1 \geq 0$ we get:

$$L(Z) = \sqrt{(x_1 + y_2)^2 + (x_2 - y_1)^2} = \sqrt{s_1^2 + s_2^2},$$

and if $x_1 y_2 - x_2 y_1 \leq 0$ we get:

$$L(Z) = \sqrt{(x_1 - y_2)^2 + (x_2 + y_1)^2} = \sqrt{t_1^2 + t_2^2}.$$

Now we write Z in its bicomplex $\mathbb{C}_i$–idempotent representation, thus we have:

$$\beta_1 = z_1 - \mathbf{i}_1 z_2 = (x_1 + y_2) + \mathbf{i}_1 (y_1 - x_2) = s_1 + \mathbf{i}_1 s_2,$$
$$\beta_2 = z_1 + \mathbf{i}_1 z_2 = (x_1 - y_2) + \mathbf{i}_1 (y_1 + x_2) = t_1 + \mathbf{i}_1 t_2,,$$

therefore

$$|\beta_1|^2 = s_1^2 + s_2^2, \qquad |\beta_2|^2 = t_1^2 + t_2^2.$$

Put together, we proved that the Lie norm of a bicomplex number is given by the following.

Proposition 7.4 *For any bicomplex number $Z = \beta_1 \mathbf{e}_{12} + \beta_2 \bar{\mathbf{e}}_{12}$ written in the $\mathbb{C}_{i_1}$–idempotent representation, its Lie norm is given by:*

$$L(Z) = \max\{|\beta_1|, |\beta_2|\}. \tag{15}$$

The dual Lie norm of bicomplex numbers is computed as follows (see [11] for the general definition in $\mathbb{C}^n$):

$$L^*(Z) := \sup\{\langle Z, W \rangle_{\mathbb{R}} \mid L(W) \leq 1\}$$
$$= \frac{\sqrt{2}}{2} \sqrt{x_1^2 + x_2^2 + y_1^2 + y_2^2 + \sqrt{(x_1^2 + x_2^2 - y_1^2 - y_2^2)^2 + 4(x_1 y_1 + x_2 y_2)^2}}$$
$$= \frac{\sqrt{2}}{2} \sqrt{x_1^2 + x_2^2 + y_1^2 + y_2^2 + \sqrt{(x_1 - y_2)^2 + (x_2 + y_1)^2} \sqrt{(x_1 + y_2)^2 + (x_2 - y_1)^2}}.$$

In idempotent coordinates, we obtain:

Proposition 7.5 *For any bicomplex number $Z = \beta_1 \mathbf{e}_{12} + \beta_2 \bar{\mathbf{e}}_{12}$, its dual Lie norm is given by:*

$$L^*(Z) = \frac{|\beta_1| + |\beta_2|}{2}. \tag{16}$$

5 Norms and Moduli on the Multicomplex Space

In the same vein, one can introduce the following norms on the multicomplex space. For the sake of brevity we will only discuss the multi-hyperbolic valued nor and the Finsler type norm which we think bring a big novelty element to the discussion. Let

$|z| = \|(x, y)\|$ be the usual Euclidean norm on $\mathbb{C}_i \simeq \mathbb{R}^2$. The open and closed disks in $\mathbb{C}_i$ (or $\mathbb{C}_j$) are denoted by $B_z(r)$ and $\bar{B}_z(r)$, for $0 < r \leq \infty$.

5.1　The Set of Multi-hyperbolic Numbers and the Multi-hyperbolic-Valued Norm

A special subalgebra of $\mathbb{BC}_n$ is the set of multi-hyperbolic numbers of one order lower d_{n-1}. A multi-hyperbolic number can be defined independently of $\mathbb{BC}_n$, generated similarly by $(n-1)$ commuting hyperbolic units over the field of real numbers. We denote by d_{n-1} the algebra of multi-hyperbolic numbers with the usual component–wise addition and multiplication. A detailed study of the multi–hyperbolic numbers is subject of a future paper, we just note that this algebra has the same 0-divisors as $\mathbb{BC}_n$.

In the multicomplex case, writing the last idempotent representation of $Z_n \in \mathbb{BC}_n$, as follows. Any $Z_n \in \mathbb{BC}_n$ can be written as:

$$Z_n = \sum_{j=1}^{2^{n-1}} \zeta_j \mathbf{e}_j,$$

where $\zeta_j \in \mathbb{C}$ and $\mathbf{e}_j \in S_{n-1}$.

Similar to the multicomplex case, $n-1$ multi-hyperbolic numbers have a unique idempotent representation with real coefficients:

$$\mathfrak{z}_{n-1} = \sum_{j=1}^{2^{n-1}} x_j \mathbf{e}_j, \tag{17}$$

where $\mathbf{e}_j$ are the same idempotents as in the multicomplex case. Define the set d_{n-1}^+ of non–negative hyperbolic numbers by:

$$d_{n-1}^+ = \left\{ \sum_{j=1}^{2^{n-1}} x_j \mathbf{e}_j \,\middle|\, x_j \geq 0 \right\}.$$

The hyperbolic algebra d_{n-1} is a subalgebra of the multicomplex numbers $\mathbb{BC}_n$. Just as in the $\mathbb{BC}$ case, $\mathbb{BC}_n$ is the algebraic closure of d_{n-1}, and it can also be seen as the complexification of d_{n-1} by using any of the imaginary units i_l.

Extending the bicomplex results, one can define a partial order relation $\preceq$ on d_{n-1} by:

$$\mathfrak{z}_{n-1_1} \preceq \mathfrak{z}_{n-1_2} \quad \text{if and only if} \quad \mathfrak{z}_{n-1_2} - \mathfrak{z}_{n-1_1} \in d_{n-1}^+. \tag{18}$$

Extending the methods in [1], one can use the partial order defined above to define a *hyperbolic-valued* norm for $\mathfrak{z}_{n-1} = \sum_{j=1}^{2^{n-1}} x_j \mathbf{e}_j$ by:

$$|\mathfrak{z}_{n-1}|_h := \sum_{j=1}^{2^{n-1}} |x_j| \mathbf{e}_j \in \mathrm{d}_{n-1}^+.$$

This extended definition obeys the corresponding properties of a norm, i.e. $|\mathfrak{z}_{n-1}|_h = 0$ if and only if $\mathfrak{z}_{n-1} = 0$, it is multiplicative: $|\mathfrak{z}_{n-1_1}\mathfrak{z}_{n-1_2}|_h = |\mathfrak{z}_{n-1_1}|_h \cdot |\mathfrak{z}_{n-1_2}|_h$, for any real number $\lambda \in \mathbb{R}$, we have $|\lambda \mathfrak{z}_{n-1}|_h = |\lambda| \cdot |\mathfrak{z}_{n-1}|_h$, and it respects the triangle inequality with respect to the order introduced above:

$$|\mathfrak{z}_{n-1_1} + \mathfrak{z}_{n-1_2}|_h \preceq |\mathfrak{z}_{n-1_1}|_h + |\mathfrak{z}_{n-1_2}|_h,$$

for all $\mathfrak{z}_{n-1_1}, \mathfrak{z}_{n-1_2} \in \mathrm{d}_{n-1}$.

For any number in d_{n-1}^+ its square root is well defined therefore, for a multicomplex number $Z_n = \sum_{j=1}^{2^{n-1}} \zeta_j \mathbf{e}_j$, one can define its conjugate by $\bar{Z}_n := \sum_{j=1}^{2^{n-1}} \bar{\zeta}_j \mathbf{e}_j$,

Definition 7.3 The *multi-hyperbolic-valued norm* is the square root of

$$Z_n \bar{Z}_n = \sum_{j=1}^{2^{n-1}} \zeta_j \bar{\zeta}_j \mathbf{e}_j = \sum_{j=1}^{2^{n-1}} \|\zeta_j\|^2 \mathbf{e}_j.$$

In the end, one obtains that the multi-hyperbolic modulus of multicomplex number is given by:

$$\|Z_n\|_h = \sum_{j=1}^{2^{n-1}} \|\zeta_j\| \mathbf{e}_j \in D_{n-1}.$$

5.2 Finsler-Type Modulus

Another real-valued 2^n–degree modulus that it is used in this setting is the one found by multiplying a multicomplex number by all its conjugates:

$$|Z_n|_{\mathscr{F}}^{2^n} := \prod_{I = i_1 \dots i_l} \bar{Z}_n^{i_1 \dots i_l}.$$

One can easily check that $|Z_n|_{\mathscr{F}}^{2^n}$ is indeed a real number, it is positive when Z_n is not a divisor of 0, and it is 0 when Z_n is singular, i.e. a complex multiple of any of the 0-divisors $\mathbf{e}_l$. These results are best seen in the idempotent representation of multicomplex numbers.

Indeed, if $Z_n = \sum_{j=1}^{2^{n-1}} \zeta_j \mathbf{e}_j$, with $\zeta_j \in \mathbb{C}_i$, then:

$$|Z_n|_{\mathscr{F}}^{2^n} = \prod_{j=1}^{2^{n-1}} |\zeta_j|^2 \in \mathbb{R}^+.$$

Note that $|Z_n|_{\mathscr{F}}^{2^n} = 0$ if and only if $\zeta_j = 0$ for some index j, i.e. Z_n is a zero divisor, or all $\zeta_j = 0$ are equal zero, i.e. $Z_n = 0$. Moreover, if $\lambda \in \mathbb{C}$ then

$$|\lambda Z_n|_{\mathscr{F}}^{2^n} = |\lambda| |Z_n|_{\mathscr{F}}^{2^n}$$

Consider $Z_n = \sum_{j=1}^{2^n 1} \zeta_j \mathbf{e}_j$ and $W_n = \sum_{j=1}^{2^{n-1}} \eta_j \mathbf{e}_j$. Then

$$|Z_n + W_n|_{\mathscr{F}}^{2^n} = \prod_j |\zeta_j + \eta_j|^2.$$

We would like to remark that the 2^n-Laplacian that appears when one considers all differential operators from all types of conjugations is best linked with this modulus and this remains an open problem.

6 Bicomplex Manifolds

The most approachable case is the case of a 4–manifold with a single bicomplex structure and we follow [3] to write this definition:

Definition 7.4 A 4-dimensional differentiable manifold M^4 is a bicomplex manifold if there is an atlas of charts locally homeomorphic to a $\mathbb{BC}$ such that the transition maps are bicomplex bi-holomorphic.

Here we use the usual sense of a bicomplex bi-holomorphic map which is a bijective map from an open in $\mathbb{BC}$ to another open in $\mathbb{BC}$, such that both the map and its inverse are bicomplex holomorphic.

One can generalize this definition in the usual way to a $4n$-dimensional manifold and we can define a general bicomplex structure as follows:

Definition 7.5 A differentiable manifold M^{4n} has a bicomplex manifold structure if there is an atlas of charts locally homeomorphic to a $\mathbb{BC}^n$ such that the transition maps are bicomplex bi-holomorphic in each component.

6.1 Examples of Bicomplex Manifolds

Following Baird and Wood [3] we describe the following examples of a one dimensional bicomplex manifold.

Example 7.1 Bicomplex Sphere The complexified sphere

$$\mathscr{S}_\mathbb{C}^2 = \{(z_1, z_2, z_3) \in \mathbb{C}^3 \mid z_1^2 + z_2^2 + z_3^2 = 1\}$$

has a one dimensional bicomplex manifold structure. In this case $\sigma_S = \dfrac{z_2}{1 + z_1} +$ $\mathbf{i}_2 \dfrac{z_3}{1 + z_1}$ is a chart from the bicomplex sphere without the north pole to the bicomplex plane without a hyperplane. In a similar way the chart from the south pole can be constructed and the transition map is bicomplex holomorphic.

There are two types of bicomplex quadrics that have a one-dimensional bicomplex structure.

Example 7.2 First Bicomplex Quadric

$$\mathbb{BC}\mathscr{Q}_1 = \{(Z_1, Z_2, Z_3) \in \mathbb{BC}^3 \mid Z_1^2 + Z_2^2 + Z_3^2 = 0, \text{ s.t. } |Z_i|_{i_2}^2 \neq 0, \text{ for some } i\}/_\simeq$$

where $(Z_1, Z_2, Z_3) \simeq (W_1, W_2, W_3)$ iff $(\lambda Z_1, \lambda Z_2, \lambda Z_3) = (W_1, W_2, W_3)$ for some $\lambda \in \mathbb{BC}_*$. In this case λ must be invertible, or (Z_1, Z_2, Z_3) or (W_1, W_2, W_3) will not meet the conditions stated.

This quadric is a one dimensional bicomplex manifold and we refer the reader to [3] for a complete description of the atlas of charts and properties of this quadric.

Example 7.3 Second Bicomplex Quadric This quadric is a subspace of $\mathbb{CP}^3$ defined as follows:

$$\mathbb{C}\mathscr{Q}_2 = \{[Z_0, Z_1, Z_2, Z_3) \in \mathbb{CP}^3 \mid Z_0^2 = Z_1^2 + Z_2^2 + Z_3^2\}$$

This quadric is a one dimensional bicomplex manifold as well, following [3].

6.2 Remarks on Bicomplex Manifolds

If one defines:

$$\mathbb{C}\mathscr{Q}_1 = \{[Z_0, Z_1, Z_2) \in \mathbb{CP}^2 \mid Z_0^2 + Z_1^2 + Z_2^2 = 0\}$$

we have that $\mathbb{C}\mathscr{Q}_1 \simeq \mathbb{CP}^1$, and $\mathbb{BC}\mathscr{Q}_1 \simeq \mathbb{C}\mathscr{Q}_1 \times \mathscr{Q}_1$ therefore $\mathbb{BC}\mathscr{Q}_1 \simeq \mathbb{CP}^1 \times \mathbb{CP}^1$ as a complex manifold. This manifold has a dense open subset $\mathbb{BC}\mathscr{Q}_1^*$ where the complex modulus of each element is non-zero, a first example of a one-dimensional manifold which is not globally a product of complex curves.

In the same way, we have that the complexified sphere is embeded in $\mathbb{C}\mathcal{Q}_2$ through the mapping $(z_1, z_2, z_3) \mapsto [1, z_1, z_2, z_3]$, which is bicomplex holomorphic. The two quadrics are equivalent as one-dimensional bicomplex manifolds and through the isomorphism $\mathbb{BC}\mathcal{Q}_1 \simeq \mathbb{C}\mathcal{Q}_2$ we have a conformal compactification of the complexified sphere.

The previous approach of Baird Wood verifies the following equivalences:

$$
\begin{array}{ccc}
\mathbb{BC} & \xrightarrow{\ Id\ } & \mathbb{BC} \\
\downarrow & & \downarrow \\
\mathbb{BC}\mathcal{Q}_1 & \xrightarrow{\ \simeq\ } & \mathbb{C}\mathcal{Q}_2 \\
{\scriptstyle 2:1}\downarrow & & {\scriptstyle 2:1}\downarrow \\
G_2(\mathbb{C}^3) & \xrightarrow{\ \perp_{\mathbb{C}}\ } & \mathbb{CP}^2
\end{array}
$$

where $G_2(\mathbb{C}^3)$ is the complex grassmanian.

7 Multicomplex Manifolds

Definition 7.6 A differentiable manifold M^{2^n} is a multicomplex manifold if there is an atlas of charts locally homeomorphic to a $\mathbb{BC}_n$ such that the transition maps are multicomplex bi-holomorphic.

Again we use the usual sense of a multicomplex bi-holomorphic map which is a bijective map from an open in $\mathbb{BC}_n$ to another open in $\mathbb{BC}_n$, such that both the map and its inverse are multicomplex holomorphic.

Remark 7.1 A 8-dimensional manifold that is multicomplex can be viewed as a bicomplex manifold but, in general an 8–dimensional manifold does not admit a multicomplex structure. This happens in all dimensions, as one needs additional Cauchy–Riemann type equations at each step.

From multicomplex analysis, one can easily obtain the following structure theorem:

Theorem 7.8 *At the idempotent level any chart is locally bi-holomorphic with a space of type:*

$$U_1 \mathbf{e}_1 + \cdots U_M \mathbf{e}_M$$

where $M = 2^{n-1}$ and U_i is either in d or in $\mathbb{C}$ for all i.

This theorem exemplifies the rigidity of a multicomplex structure as well as its ability of combining a theory of several complex variables into a single multicomplex variable.

In the examples that follow we will define two types of multicomplex spheres, each following the two dimensional example in its own way. For each example we will make use of the *nesting* definition of the multicomplex spaces, but in different ways.

Example 7.4 (*First Multicomplex Sphere*) One can define a multicomplex structure on the following object:

$$\mathbb{BC}_n \mathscr{S}_1 = \{(Z_{n-1,1}, Z_{n-1,2}, Z_{n-1,3}) \in \mathbb{BC}_{n-1}^3 \mid Z_1^2 + Z_2^2 + Z_3^2 = 1\},$$

as an example of a chart we can take:

$$\sigma_{n,N}(Z_{n-1,1}, Z_{n-1,2}, Z_{n-1,3}) = \frac{Z_{n-1,2}}{1 + Z_{n-1,1}} + \mathbf{i}_n \frac{Z_{n-1,3}}{1 + Z_{n-1,1}}$$

defined on the space of triplets:

$$\{(Z_{n-1,1}, Z_{n-1,2}, Z_{n-1,3}) \in \mathbb{BC}_{n-1}^3 \mid Z_{n-1,1}^2 + Z_{n-1,2}^2 + Z_{n-1,3}^2 = 1\} \backslash$$
$$\backslash \{(1 + Z_{n-1,1}) \text{ invertible}\}.$$

In a similar way one can define the other charts and the reader can check that the transition functions are multicomplex holomorphic.

Example 7.5 ($\mathbb{BC}_3$ *Sphere of the second type*) Our first example will be a $\mathbb{BC}_3$ structure on the 4-dimensional complexified sphere:

$$\mathscr{S}_{\mathbb{C}}^4 = \{(z_0, z_1, z_2, z_3, z_4) \in \mathbb{C}^5 \mid z_0^2 + z_1^2 + z_2^2 + z_3^2 + z_4^3 = 1\}$$

This sphere can be endowed with an $\mathbb{BC}_3$ structure via the following type of charts:

$$\sigma_{3,N}(z_0, z_1, z_2, z_3, z_4) = \left(\frac{z_1}{1 + z_0} + \mathbf{i}_2 \frac{z_2}{1 + z_0}\right) + \mathbf{i}_3 \left(\frac{z_3}{1 + z_0} + \mathbf{i}_2 \frac{z_4}{1 + z_0}\right)$$

defined on the space:

$$\{(z_0, z_1, z_2, z_3, z_4) \in \mathbb{C}^5 \mid z_0^2 + z_1^2 + z_2^2 + z_3^2 + z_4^3 = 1\} \backslash \{(z_0 \neq -1)\}.$$

In a similar way one can define the other charts and the reader can check that the transition functions are $\mathbb{BC}_3$ holomorphic.

Example 7.6 (*Multicomplex Sphere of the second type*) One can define a $\mathbb{BC}_n$ multicomplex structure on the 2^{n-1}-dimensional complexified sphere as well, through an inductive process.

$$\mathscr{S}_{\mathbb{C}}^{2^{n-1}+1} = \{(z_0, z_1, \ldots, z_{2^{n-1}}) \in \mathbb{C}^{2^{n-1}+1} \,|\, z_0^2 + z_1^2 + \cdots + z_{2^{n-1}}^2 = 1\}$$

This sphere can be endowed with an $\mathbb{BC}_n$ structure through a set of *nested* charts, and we will only write the $\mathbb{BC}_4$ one to give an idea of the inductive process.

$$\sigma_{4,N}(z_0, z_1, \ldots, z_8) = \left(\left(\frac{z_1}{1+z_0} + \mathbf{i}_2\frac{z_2}{1+z_0}\right) + \mathbf{i}_3\left(\frac{z_3}{1+z_0} + \mathbf{i}_2\frac{z_4}{1+z_0}\right)\right) \quad (19)$$

$$+ \mathbf{i}_4\left(\left(\frac{z_5}{1+z_0} + \mathbf{i}_2\frac{z_6}{1+z_0}\right) + \mathbf{i}_3\left(\frac{z_7}{1+z_0} + \mathbf{i}_2\frac{z_8}{1+z_0}\right)\right)$$
$$(20)$$

defined on the space:

$$\{(z_0, \ldots, z^{2^{n-1}}) \in \mathbb{C}^{2^{n-1}+1} \,|\, z_0^2 + z_1^2 + \cdots + z_{2^{n-1}}^2 = 1\} \setminus \{(z_0 \neq -1)\}.$$

In a similar way one can define the other charts and the reader can check that the transition functions are $\mathbb{BC}_n$ holomorphic.

It is worth noting that the bicomplex sphere (the complexified sphere) has a natural embedding in both of these types of multicomplex spheres, in the first case as intersection with hyperplanes, in the second as a "nested" submanifold theory of higher and higher dimension.

This is a first example of a submanifold theory in multicomplex manifold theory, exemplifying the complexity of this case.

8   Conclusions and Future Endeavors

In this paper we are opening new avenues of discussion and ideas which can be applied in all commutative settings, for example the ternary algebra [2], as well as other hypercomplex algebras, especially ones coming from quotients of polynomial rings. We are working towards studying submanifold theory coming from the study of harmonic maps in this context and we think that these methods may apply to more areas.

I would also like to mention other approaches to the theory of bicomplex manifolds, as we would be remiss if we did not mention that a bicomplex structure can be defined in a different way, reminiscing of the definition of quaternionic manifolds in [15]. This definition does not use the concepts of bicomplex and multicomplex holomorphic maps, but it is worth noting some results in this direction.

Following Salamon's definition of a quaternionic manifold, in [8, 9] the authors have defined the notion of almost product bicomplex structure on a manifold as follows.

Definition 7.7 An almost product bicomplex (apbc)-structure on a paracompact connected C^∞-manifold M is a triple (F, G, H) of $(1, 1)$-tensor fields which satisfies the conditions

$$-F^2 = G^2 = H^2 = F \circ G \circ H = -I, \quad F \neq \pm I. \tag{21}$$

It follows that F is an almost product (ap)–structure and G, H are almost complex (ac)–structures on M, which satisfy the relations:

$$F \circ G = G \circ F = H, \quad G \circ H = H \circ G = -F,$$
$$H \circ F = F \circ H = G, \quad F \neq \pm I.$$

In this context, the authors prove several structure theorems, as well as a very interesting integrability theorem and it is worth noting that our bicomplex manifolds inherit an apbc structure from their bicomplex ambient spaces.

Another direction comes from the fact that the bicomplex manifolds defined in this paper are intrinsically (apbc) in the sense of Sect. 6. A following paper will investigate the integrability of the structures in the examples given as well as the relationship between the holomorphicity of the structure maps and the integrability of the (apbc) manifold.

We are currently developing the geometry of multicomplex manifolds endowed with Riemann–Finsler metrics. As we saw above, we have a non-canonical metric on both the bicomplex and the multicomplex space. In other works we can see that this metric arises naturally from the structure of the multicomplex space as a quotient polynomial ring.

It is of great interest to fully understand a submanifold theory in these ambient spaces, as well as the relationship between Riemannian–Finsler metrics on these spaces at each level. These issues are currently under investigation and we hope to show the relationship between different types of submanifolds in these spaces and a harmonic map theory in each case.

References

1. Alpay, D., Luna-Elizarrarás, M.E., Shapiro, M., Struppa, D.C.: Basics of functional analysis with bicomplex scalars, and bicomplex Schur analysis. Springer Briefs in Mathematics. Springer, Cham (2014)
2. Alpay, D., Vajiac, A., Vajiac, M.: Gleason's problem associated to a real ternary algebra and applications. Adv. Appl. Clifford Algebras **28**, 1–16 (Springer International Publishing)
3. Baird, P., Wood, J.C.: Harmonic morphisms and bicomplex manifolds. J. Geom. Phys. **61**(1), 46–61 (2011)

4. Catoni, F., Boccaletti, D., Cannata, R., Catoni, V., Nichelatti, E., Zampetti, P.: The mathematics of Minkowski space–time. Birkhäuser, Basel (2008)
5. Charak, K.S., Rochon, D., Sharma, N.: Normal families of bicomplex holomorphic functions. arXiv:0806.4403v1 (2008)
6. Colombo, F., Sabadini, I., Struppa, D.C., Vajiac, A., Vajiac, M.: Singularities of functions of one and several bicomplex functions. Ark. Mat. **49**, 277–294 (2011)
7. Colombo, F., Sabadini, I., Struppa, D.C., Vajiac, A., Vajiac, M.: Bicomplex hyperfunction theory. Ann. Mat. Pura Appl. **190**, 247–261 (2011)
8. Cruceanu, V.: Almost product bicomplex structures on manifolds. Analele Stiintifice ale Univ. Al.I.Cuza, Iasi, Tomul LI, s.I, Matematica (2005) (f.1)
9. Cruceanu, V., Fortuny, P., Gadea, P.: A survey on paracomplex geometry. Rocky Mt. J. Math **26**(1), 83–115 (1996)
10. Luna-Elizarraras, M.E., Shapiro, M., Struppa, D.C., Vajiac, A.: Bicomplex Holomorphic Functions: The Algebra, Geometry and Analysis of Bicomplex Numbers. Frontiers in Mathematics. Birkhäuser, (2016)
11. Morimoto, M., Fujita, K.: Between lie norm and dual lie norm. Tokyo J. Math. **24**(2), 499–507 (2001)
12. Price, G.B.: An Introduction to Multicomplex Spaces and Functions. Marcel Dekker, New York (1991)
13. Ryan, J.: Complexified Clifford analysis. Complex Var. Elliptic Equ. **1**, 119–149 (1982)
14. Sabadini, I., Vajiac, A., Vajiac, M.: Bernstein-type inequalities for bicomplex polynomials. In: Advances in Complex Analysis and Operator Theory, pp. 281–299. Birkhäuser, Cham (2017)
15. Salamon, S.M.: Differential geometry of Quaternionic manifolds. Annales sci. de l'École Normale Supérieure, Sér. 4. **19**(1), 31–55 (1986)
16. Scorza-Dragoni, G.: Sulle funzioni olomorfe di una variabile bicomplessa. Reale Accad. d'Italia, Mem. Classe Sci. Nat. Fis. Mat. **5**, 597–665 (1934)
17. Segre, C.: Le: rappresentazioni reali delle forme complesse e gli enti iperalgebrici. Math. Ann. **40**, 413–467 (1892)
18. Sommen, F.: Spherical monogenics on the lie sphere. J. Func. Anal. **92**, 372–402 (1990)
19. Spampinato, N.: Estensione nel campo bicomplesso di due teoremi, del Levi-Civita e del Severi, per le funzioni olomorfe di due variabili bicomplesse I, II. Reale Accad. Naz. Lincei **22**(38–43), 96–102 (1935)
20. Spampinato, N.: Sulla rappresentazione di funzioni di variabile bicomplessa totalmente derivabili. Ann. Mat. Pura Appl. **14**, 305–325 (1936)
21. Struppa, D.C., Vajiac, A., Vajiac, M.: Remarks on Holomorphicity in three settings: complex, quaternionic, and bicomplex. In: Hypercomplex Analysis and Applications. Trends in Mathematics, pp. 261–274. Birkhäuser, Basel (2010)
22. Struppa, D.C., Vajiac, A., Vajiac, M.: Holomorphy in multicomplex spaces, spectral theory, mathematical system theory, evolutions equations, differential and difference equations. In: Operator Theory: Advances and Applications, vol. 221, pp. 617–634. Birkhäuser, Basel (2012)
23. Struppa, D.C., Vajiac, A., Vajiac, M.: Differential equations in multicomplex spaces, hypercomplex analysis: new perspectives and applications. In: Trends in Mathematics, pp. 213–227 (2014)
24. Vajiac, A., Vajiac, M.: Multicomplex hyperfunctions. Complex Var. Elliptic Equ. (2011). https://doi.org/10.1080/17476933.2011.603419

Associated Operators to the Space of Meta-q-Monogenic Functions

C. J. Vanegas and F. A. Vargas

Abstract We are giving a characterization of all linear first order partial differential operators with Clifford-algebra-valued coefficients that are associated to the meta-q-monogenic operator. As an application, the solvability of initial value problems involving these operators is shown.

Keywords Associated spaces · Meta-q monogenic functions · Clifford algebras

AMS Classification Primary 35F10 · Secondary 35A10, 15A66

1 Introduction

Let $\{e_0, e_1, \ldots, e_n\}$ be an orthonormal basis of the Euclidean space $\mathbb{R}^{n+1}$. Let $\mathscr{A}_n$ be the Clifford algebra defined as the 2^n-dimensional real associative and non-commutative algebra, whose basis is denoted by

$$\beta = \{e_N : N \in \Gamma_n\}, \quad \Gamma_n = \{0, 1, 2, \ldots, 12, 13, \ldots, 123 \ldots n\}.$$

In [7] a product on $\mathscr{A}_n$ is defined by

$$e_A \cdot e_B = (-1)^{n(A \cap B)}(-1)^{p(A,B)} e_{A \triangle B}, \tag{1}$$

C. J. Vanegas (✉) · F. A. Vargas
Departamento de Matemáticas y Estadísticas, ICB, Universidad Técnica de Manabí,
Portoviejo, Ecuador
e-mail: cvanegas@utm.edu.ec; cvanegas@usb.ve

C. J. Vanegas
Departamento de Matemáticas Puras Y Aplicadas, Universidad Simón Bolívar,
Caracas, Venezuela

F. A. Vargas
Departamento de Cómputo Científico y Estadística, Universidad Simón Bolívar,
Caracas, Venezuela
e-mail: franklinvj@usb.ve; favargas@utm.edu.ec

© Springer Nature Switzerland AG 2018
P. Cerejeiras et al. (eds.), *Clifford Analysis and Related Topics*,
Springer Proceedings in Mathematics & Statistics 260,
https://doi.org/10.1007/978-3-030-00049-3_8

141

where $n(A)$ is the cardinal of the set A, i. e., $n(A) = \#A$, and $p(A, B) = \sum_{j \in B} p(A, j)$,

where $p(A, j) = \#\{i \in A : i > j\}$. The sets A, B and $A \triangle B$ (symmetric difference of A and B) are ordered in the prescribed way. It follows from the multiplication rule (1) that e_0 is the identity element, $e_i e_j + e_j e_i = -2\delta_{ij}$, where δ_{ij} is the Kronecker symbol and $e_{h_1} \cdot e_{h_2} \cdots e_{h_r} = e_{h_1 \ldots h_r}$ for $1 \leq h_1 < \cdots < h_r \leq n$.

Let Ω be an open set in $\mathbb{R}^{n+1}$. The $\mathscr{A}_n$-valued functions u defined in Ω, can be written as

$$u(x) = \sum_{N \in \Gamma_n} u_N(x) e_N, \quad x = (x_0, \ldots, x_n) \in \Omega, \tag{2}$$

where each u_N is a real-valued function. We say that these functions $u = u(x)$ have the properties of continuity, differentiability, etc. if each its components $u_N = u_N(x)$, $N \in \Gamma_n$ has these properties.

The $\mathscr{A}_n$-valued function u, defined and twice continuously differentiable in Ω, is called a monogenic function in Ω if $\mathscr{D}u = 0$, where $\mathscr{D} = \sum_{i=0}^n e_i \partial_i$, where ∂_i is the differentiation operator with respect to x_i, is the generalized Cauchy–Riemann operator.

The concept of associated spaces comes from complex analysis: The space of holomorphic functions is associated to the complex differentiation $\dfrac{d}{dz}$ because the complex derivative of a holomorphic function is again holomorphic. Generalizing this idea, we say that a function space $\mathscr{X}$ is called an associated space to a given differential operator $\mathscr{F}$ if $\mathscr{F}$ transform $\mathscr{X}$ into itself [13, 15].

Associated spaces are used to solve initial value problems of the type

$$\partial_t u = \mathscr{F}(t, x, u, \partial_j u), \quad j = 0, \ldots, n, \tag{3}$$
$$u(0, x) = \varphi(x), \tag{4}$$

where $\varphi(x)$ satisfies the partial differential equation $\mathscr{G}(u) = 0$, provided that the initial function $\varphi(x)$ belongs to the associated space $\mathscr{X}$ of $\mathscr{F}$ containing all the solutions for $\mathscr{G}(u) = 0$, and that the elements of $\mathscr{X}$ satisfy an interior estimate, i.e., an estimate for the derivatives of the solutions near the boundary of a certain bounded domain. Although the operator $\mathscr{F}$ depends on the spacelike variable x and on the time t, the elements of an associated space must be functions depending only on the same spacelike variable x and not additionally on t.

We observe that the solutions of the initial value problem (3), (4) are fixed points of the integro-differential operator

$$Tu(t, x) = \varphi(x) + \int_0^t \mathscr{F}(\tau, x, u(\tau, x), \partial_j u(\tau, x)) d\tau. \tag{5}$$

and vice versa (see [9]). In order to apply a fixed-point theorem such as the contraction mapping principle, the operator (5) has to be estimated in a suitable function space. As $\mathscr{F}(\tau, x, u, \partial_j u)$ in (5) also depends on the derivatives $\partial_j u$, the operator (5) maps a certain Banach function space into itself in case the derivatives $\partial_j(Tu(t, x))$ do exist and can be estimated in a suitable way, see [14, 15].

There are two basic problems in the theory of associated spaces. The first one is the direct problem, which consists on the construction of an associated space $\mathscr{X}$ to a given operator $\mathscr{F}$, whereas the second one is the inverse problem, dealing on finding an operator $\mathscr{F}$ defined on a given space $\mathscr{X}$ such that $\mathscr{X}$ is associated to $\mathscr{F}$ [15].

In this work we are focused on an inverse problem: We consider the linear first order partial differential operator $\mathscr{F}$ defined by

$$\mathscr{F}u = \sum_{i=0}^{n} A^{(i)}(x)\partial_i u(x) + B(x)u(x) + C(x), \qquad (6)$$

where $x = (x_0, \ldots, x_n)$ is a spacelike variable on $\mathbb{R}^{n+1}$ and $u = u(x)$ is an $\mathscr{A}_n$-Clifford algebra-valued function. The coefficients $A^{(i)}$ are $\mathscr{A}_n$-valued and twice continuously differentiable functions and B and C are $\mathscr{A}_n$-valued and continuously differentiable functions. Then we will determine necessary and sufficient conditions such that the operator $\mathscr{F}$ be associated to the meta-q-monogenic operator

$$\mathscr{D}_{(q,\lambda)} = \sum_{i=0}^{n} q_i \partial_i + \lambda, \qquad (7)$$

where $q_0 = 1$, $q_i \in \mathscr{A}_n$, $i = 0, 1, 2, \ldots, n$ are constants and $\lambda \in \mathbb{R}$, be an associated pair of operators. Observe that if we take $q_i = e_i$ and $\lambda = 0$, we recover the generalized Cauchy–Riemann operator $\mathscr{D}$.

Sufficient conditions for evolution operators transforming regular or monogenic functions into themselves are given in [5, 10, 12, 16, 17], in the framework of quaternionic analysis and Clifford analysis, respectively.

Necessary and sufficient conditions in quaternionic analysis are given in [1–3, 11], and recently the works [4, 6] appeared showing necessary and sufficient conditions in Clifford analysis.

The results in our paper provide a generalization of all those mentioned above. As an application, we show the solvability of initial value problems involving the operators $\mathscr{F}$ and $\mathscr{D}_{(q,\lambda)}$.

2 The Meta-q-Monogenic Operator

We consider the meta-q-monogenic operator defined by (7) and the associated system

$$\mathscr{D}_q u = \sum_{i=0}^{n} q_i \partial_i u = 0. \qquad (8)$$

This system is not necessarily a Cauchy–Riemann system. To fulfill this property we have

Theorem 8.1 *The Eq. (8) is a Cauchy–Riemann system if and only if the coefficients* q_i *satisfy the relations*

$$q_i \overline{q_j} + q_j \overline{q_i} = 2\delta_{ij}, \quad i, j = 0, 1, 2, \ldots, n. \tag{9}$$

Note that the second order operators obtained from $\mathcal{D}_q$ and its conjugate are

$$\mathcal{D}_q \overline{\mathcal{D}_q} = \sum_{i=0}^{n} q_i \overline{q_i} \partial_i^2 + \sum_{i<j} (q_i \overline{q_j} + q_j \overline{q_i}) \partial_i \partial_j \tag{10}$$

and

$$\overline{\mathcal{D}_q} \mathcal{D}_q = \sum_{i=0}^{n} \overline{q_i} q_i \partial_i^2 + \sum_{i<j} (\overline{q_i} q_j + \overline{q_j} q_i) \partial_i \partial_j. \tag{11}$$

If we take the parameters q_i arbitrarily in $\mathscr{A}_n$, then (10) and (11) are not scalar operators. But if the parameters q_i satisfy the structure relations (9), then we get the Laplacian, i.e.,

$$\overline{\mathcal{D}_q} \mathcal{D}_q = \mathcal{D}_q \overline{\mathcal{D}_q} = \Delta_{n+1}.$$

In general, if we take an orthonormal basis $\{q_1, q_2, \ldots, q_n\}$ of $\mathbb{R}^n$ other than standard base and $q_0 = 1$, then the former relations are satisfied. Operators $\mathcal{D}_q$ with a basis other than the usual to generate the algebra, were studied by [2] in the quaternionic case, from where we have the following extended definitions to $\mathscr{A}_n$:

Definition 8.1 A set $q = \{q_0, q_1, \ldots, q_n\} \subset \mathscr{A}_n$ is said to be a structural set if it satisfies the relations (9).

Definition 8.2 Two structural sets p, q are said to be left equivalent (resp. right) if there exists $h \in \mathscr{A}_n$, $|h| = 1$ such that

$$p = h \cdot q \quad (\text{resp. } p = q \cdot h).$$

The last definition represents an equivalence class on the collection of structured sets. Moreover, each equivalence class has a unique representative structural set q with $q_0 = 1$.

The following properties can be obtained from the structural relations (9).

Proposition 8.1 *Let q be a structural set with $q_0 = 1$ in $\mathscr{A}_n$. Then for $i = 1, 2, \ldots, n$*

1. $\overline{q_i} = -q_i$
2. $q_i^2 = -1$
3. $q_i q_j + q_j q_i = 0, i \neq j.$

In the quaternionic case, the elements of a structural set are necessarily vectors. In $\mathscr{A}_n$ we also can construct structural set whose elements are vectors. If $\{b_1, b_2, \ldots, b_n\}$ is an orthonormal basis of $\mathbb{R}^n$, where the components of each vector b_i are denoted by b_{ij}, $i, j = 1, \ldots, n$, then the set

$$q_0 = 1 \quad \text{and} \quad q_i = \sum_{j=1}^{n} b_{ij} e_j, \quad i = 1, 2, \ldots, n, \tag{12}$$

forms a structural set in $\mathscr{A}_n$. One question that should be answered is whether there are structural sets in $\mathscr{A}_n$, whose elements are not necessarily vectors.

The first property in Proposition 8.1 implies that the explicit construction of q_i satisfying the structure relations (9) will depend on the selected involution for $\mathscr{A}_n$. For example the involution

$$\overline{e_N} = (-1)^{|N|} e_N, \quad N \in \Gamma, \tag{13}$$

implies that the elements q_i must belong to the odd subalgebra of $\mathscr{A}_n$. We conjecture that a structural set in this subalgebra necessarily have to be of the form (12).

3 Associated Spaces

Definition 8.3 ([13, 15]) Let $\mathscr{F}$ be a first order differential operator depending on t, x, u and $\partial_i u$ for $i = 0, 1, \ldots, n$, while $\mathscr{G}$ is a differential operator with respect to the spacelike variables x_i with coefficients not depending on time t. $\mathscr{F}$ is said to be associated with $\mathscr{G}$ if $\mathscr{F}$ maps solutions for the differential equation $\mathscr{G}u = 0$ into solutions of the same equation for a fixedly chosen t, i.e.,

$$\mathscr{G}u = 0 \Rightarrow \mathscr{G}(\mathscr{F}u) = 0.$$

The function space $\mathscr{X}$ containing all the solutions for the differential equation $\mathscr{G}u = 0$ is called an associated space of $\mathscr{F}$.

Next we will determine necessary and sufficient conditions such that the operator $\mathscr{F}$ defined by (6) be associated to the meta-q-monogenic operator (7).

3.1 Necessary and Sufficient Conditions on the Coefficients of $\mathscr{F}$

We consider the operator $\mathscr{F}$ defined by (6) and we will determine conditions over $A^{(i)}$, B and C guaranteeing

$$\mathscr{D}_{(q,\lambda)}u = 0 \Rightarrow \mathscr{D}_{(q,\lambda)}(\mathscr{F}u) = 0. \tag{14}$$

Let u and v be continuously differentiable functions defined in $\mathbb{R}^{n+1}$ and taking values in $\mathscr{A}_n$. Is easy to see that the following identity is true [5]

$$\partial_i(u \cdot v) = \partial_i u \cdot v + u \cdot \partial_i v, \quad i = 0, \dots, n. \tag{15}$$

Since $\mathscr{D}_{(q,\lambda)}u = 0$, then

$$\partial_0 u = - \sum_{i=1}^{n} q_i \partial_i u - \lambda u. \tag{16}$$

Taking the derivative of this equation with respect to x_j we get

$$\partial_j \partial_0 u = - \sum_{i=1}^{n} \partial_j q_i \partial_i u - \sum_{i=1}^{n} q_i \partial_i \partial_j u - \partial_j \lambda u - \lambda \partial_j u, \quad j = 1, 2, \dots, n. \tag{17}$$

Substituting the expression given by (16) into $\mathscr{F}$ we obtain

$$\mathscr{F}u = \sum_{i=1}^{n} H^{(i)} \cdot \partial_i u + G \cdot u + C,$$

where

$$H^{(i)} = A^{(i)} - A^{(0)}q_i, \quad i = 0, 1, 2, \dots, n \quad \text{y} \quad G = B - A^{(0)}\lambda.$$

On the one hand we have,

$$\mathscr{D}_{(q,\lambda)}\left(\sum_{j=1}^{n} H^{(j)} \cdot \partial_j u \right) = \sum_{j=1}^{n} \left(\partial_0 H^{(j)} + \sum_{i=1}^{n} q_i \partial_i H^{(j)} + \lambda H^{(j)} \right) \cdot \partial_j u$$

$$+ \sum_{i,j=1}^{n} q_i H^{(j)} \cdot \partial_i \partial_j u + \sum_{j=1}^{n} H^{(j)} \cdot \partial_0 \partial_j u.$$

Using the Eq. (17) in the last expression we obtain

$$\mathscr{D}_{(q,\lambda)}\left(\sum_{j=1}^{n}H^{(j)}\cdot\partial_j u\right) = \sum_{j=1}^{n}\left(\mathscr{D}_{(q,\lambda)}H^{(j)} - \sum_{i=1}^{n}H^{(i)}\partial_i q_j - H^{(j)}\lambda\right)\partial_j u$$

$$+ \sum_{i,j=1}^{n}\left(q_i H^{(j)} - H^{(j)}q_j\right)\cdot\partial_i\partial_j u$$

$$- \left(\sum_{j=1}^{n}H^{(j)}\partial_j\lambda\right)u$$

$$= \sum_{j=1}^{n}\left(\mathscr{D}_{(q,\lambda)}(H^{(j)}) - \mathscr{D}_{(H,0)}(q_j) - H^{(j)}\lambda\right)\cdot\partial_j u$$

$$+ \sum_{i,j=1}^{n}\left(q_i H^{(j)} - H^{(j)}q_i\right)\cdot\partial_i\partial_j u - \mathscr{D}_{(H,0)}(\lambda)\cdot u.$$

On the other hand,

$$\mathscr{D}_{(q,\lambda)}(G\cdot u + C) = \mathscr{D}_{(q,\lambda)}(G\cdot u) + \mathscr{D}_{(q,\lambda)}(C)$$

$$= \partial_0(G\cdot u) + \sum_{i=1}^{n}q_i\partial_i(G\cdot u) + \lambda G\cdot u + \mathscr{D}_{(q,\lambda)}(C)$$

$$= (\mathscr{D}_{(q,\lambda)}(G) - G\lambda)\cdot u + \sum_{i=1}^{n}(q_i G - Gq_i)\cdot\partial_i u$$

$$+ \mathscr{D}_{(q,\lambda)}(C).$$

Adding the two previous expressions we obtain

$$\mathscr{D}_{(q,\lambda)}\mathscr{F}u = \sum_{i=1}^{n}R_i\partial_i u + \sum_{i=1}^{n}P_i\partial_i^2 u + \sum_{i<j}^{n}Q_{ij}\partial_i\partial_j u + S\cdot u + \mathscr{D}_{(q,\lambda)}(C), \quad (18)$$

where

$$R_i = \mathscr{D}_{(q,\lambda)}(H^{(i)}) - \mathscr{D}_{(H,G)}(q_i) - H^{(i)}\lambda + q_i G,$$
$$P_i = q_i H^{(i)} - H^{(i)}q_i,$$
$$Q_{ij} = q_i H^{(j)} - H^{(j)}q_i + q_j H^{(i)} - H^{(i)}q_j,$$
$$S = \mathscr{D}_{(q,\lambda)}(G) - \mathscr{D}_{(H,G)}(\lambda),$$

for $i, j = 1, 2, \ldots, n$, $i < j$. Using $H^{(i)} = A^{(i)} - A^{(0)}q_i$ and $G = B - A^{(0)}\lambda$ in these equalities, we get

$$R_i = \mathscr{D}_{(q,\lambda)}(A^{(i)} - A^{(0)}q_i) - \mathscr{D}_{(H,G)}(q_i) - (A^{(i)} - A^{(0)}q_i)\lambda$$
$$+ q_i(B - A^{(0)}\lambda),$$
$$P_i = q_i(A^{(i)} - A^{(0)}q_i) - (A^{(i)} - A^{(0)}q_i)q_i,$$
$$Q_{ij} = q_i(A^{(j)} - A^{(0)}q_j) - (A^{(j)} - A^{(0)}q_j)q_i + q_j(A^{(i)} - A^{(0)}q_i)$$
$$- (A^{(i)} - A^{(0)}q_i)q_j,$$
$$S = \mathscr{D}_{(q,\lambda)}(B - A^{(0)}) - \mathscr{D}_{(H,G)}(\lambda).$$

We observe that if

$$R_i = P_i = Q_{ij} = S = \mathscr{D}_{(q,\lambda)}(C) = 0 \quad \text{for } i, j = 1, 2, \ldots, n \text{ with } i < j, \quad (19)$$

are met then $\mathscr{D}_{(q,\lambda)}\mathscr{F}u = 0$ if $\mathscr{D}_{(q,\lambda)}u = 0$.

Now we assume that $(\mathscr{F}, \mathscr{D}_{(q,\lambda)})$ is an associated pair, i.e., $\mathscr{D}_{(q,\lambda)}(\mathscr{F}u) = 0$ if only u is a meta-q-monogenic function. In order to obtain the conditions on the coefficients of operator $\mathscr{F}$ we will start by choosing special functions of the associated space, in this case special meta-q-monogenic functions, and we will write out the relations assuming that $\mathscr{F}$ is meta-q-monogenic for those functions.

Choosing the meta-q-monogenic function $u = 0$ we have $\mathscr{D}_{(q,\lambda)}C = 0$ and so C is meta-q-monogenic and the term in $\mathscr{D}_{(q,\lambda)}C$ can be omitted. We now choose the meta-q-monogenic function $u = e^{-\lambda x_0}$ to get $\mathscr{D}_{(q,\lambda)}(\mathscr{F}u) = S = 0$. Next we choose the meta-q-monogenic functions $u = (x_j - x_0 q_j)e^{-\lambda x_0}, j = 1, 2, \ldots, n$. For these functions we obtain $\mathscr{D}_{(q,\lambda)}(\mathscr{F}u) = R_j e^{-\lambda x_0} = 0$, then $R_j = 0$ for each $j = 1, \ldots n$. Choosing the meta-q-monogenic functions $u = (x_j^2 + x_0^2 q_j^2 - 2x_0 x_j q_j)e^{-\lambda x_0}, j = 1, 2, \ldots, n$, we have $\mathscr{D}_{(q,\lambda)}(\mathscr{F}u) = 2P_j e^{-\lambda x_0} = 0$. Hence $P_j = 0$. Finally for the meta-q-monogenic function $u = (\omega_i q_j \omega_i q_j x_i^2 + x_j^2 - 2\omega_i q_j x_i x_j)e^{-\lambda x_0} i, j = 1, 2, \ldots, n, i < j$, where $\omega_i \in \mathscr{A}_n$ such that $q_i \omega_i = e_0$, we get $\mathscr{D}_{(q,\lambda)}(\mathscr{F}u) = -2Q_{ij}\omega_i q_j e^{-\lambda x_0}$. Therefore $Q_{ij} = 0$ for $i, j = 1, 2, \ldots, n$ $i < j$.

Therefore the following statement is true:

Theorem 8.2 *Let u be a $\mathscr{A}_n$-valued and twice continuously differentiable function. Suppose that $q_i \in \mathscr{A}_n$ are constants different from zero and $\lambda \in \mathbb{R}$. Then the operator $\mathscr{F}$ defined by (6) is associated to the $\mathscr{D}_{(q,\lambda)}$ operator defined by (7) if and only if the conditions (19) are satisfied.*

The special case $n = 1$

For reasons of a simpler calculation, we show the Theorem 8.2 just in the case $\mathscr{A}_1$. If $n = 1$ the conditions (19) reduce to

$$\partial_0 H^{(1)} + q_1 \partial_1 H^{(1)} = -q_1 G, \tag{20}$$

$$\partial_0 G + q_1 \partial_1 G = -\lambda G, \tag{21}$$

$$q_1 H^{(1)} - H^{(1)} q_1 = 0. \tag{22}$$

The last condition is satisfied because $\mathscr{A}_1$ is a commutative algebra. If $q_1 \neq 0$ we can solve for G from the first condition and then use it in the second one. So we have

$$\partial_0(q_1^{-1}\partial_0 H^{(1)} + \partial_1 H^{(1)}) + q_1\partial_1(q_1^{-1}\partial_0 H^{(1)} + \partial_1 H^{(1)})$$
$$+ \lambda(q_1^{-1}\partial_0 H^{(1)} + \partial_1 H^{(1)}) = 0,$$

equivalently

$$\partial_0^2 H^{(1)} + 2q_1\partial_0\partial_1 H^{(1)} + q_1^2\partial_1^2 H^{(1)} + \lambda(\partial_0 H^{(1)} + q_1\partial_1 H^{(1)}) = 0$$

i.e., $H^{(1)}$ should be solution of the second order equation:

$$\mathscr{D}_{(q,\lambda)}\mathscr{D}_q u = 0.$$

4 Application of Associated Spaces to Initial Value Problems

We consider the initial value problem

$$\partial_t u(t, x) = \mathscr{F}(t, x, u, \partial_j u) \tag{23}$$
$$u(0, x) = \varphi(x), \tag{24}$$

where $t \in [0, T]$ is the variable time, $x = (x_0, \ldots, x_n)$ is a spacelike variable on $\mathbb{R}^{n+1}$ and $u = u(t, x)$ is a Clifford algebra-valued function. The right-hand side operator $\mathscr{F}$ is taken as

$$\mathscr{F}(t, x, u, \partial_j u) = \sum_{i=0}^{n} A^{(i)}(t, x)\partial_i u(t, x) + B(t, x)u(t, x) + C(t, x), \tag{25}$$

where $A^{(i)}$ are Clifford algebra-valued and twice continuously differentiable functions and B and C are Clifford algebra-valued and continuously differentiable functions, for each $t \in [0, T]$. Furthermore, we assume that the initial function is a meta-q-monogenic function defined in a certain bounded domain Ω in $\mathbb{R}^{n+1}$.

We will show that this initial value problem is solvable if the meta-q-operator $\mathscr{D}_{(q,\lambda)}$ is associated to the operator $\mathscr{F}$ given by (25) and its elements satisfy an interior estimate.

4.1 Solvability of Initial Value Problems

Now we show the solvability of initial value problems using the theory of associated spaces, see [15] and the references therein.

The initial value problem (23), (24) can be rewritten as [9]

$$u(t, x) = \varphi(x) + \int_0^t \mathscr{F}(\tau, x, u(\tau, x), \partial_j u(\tau, x)) d\tau. \tag{26}$$

Consequently, the solution of the initial value problem (23), (24) is a fixed point of the operator

$$Tu(t, x) = \varphi(x) + \int_0^t \mathscr{F}(\tau, x, u(\tau, x), \partial_j u(\tau, x)) d\tau. \tag{27}$$

and vice versa. To apply a fixed point theorem as the Contraction Mapping Principle, the operator (27) should map a certain Banach space B of meta-q-monogenic functions into itself. Since the operator $\mathscr{F}$ also depends on the derivatives $\partial_{x_i} u$, that map exists in case the derivatives $\partial_{x_i}(Tu(t, x))$ do exist and can be estimated accordingly. Therefore, one has to restrict the operator to a space of the meta-q-monogenic functions for which the derivatives $\partial_{x_i} u$ of a meta-q-monogenic function u can be estimated by u itself. This space is the so-called associated space and the estimates for the derivatives $\partial_{x_i} u$ can be attained by using the so-called interior estimate. The necessary interior estimate (in some norms) can be achieved from integral representations obtained using fundamental solutions. They have the form:

$$\|\partial_i u\|_K \le \frac{const.}{dist(K, \partial \Omega)} \|u\|_\Omega , \tag{28}$$

where K is a compact subset of $\Omega \in \mathbb{R}^{n+1}$.

In consequence, we have the following theorem:

Theorem 8.3 *Let $\mathscr{F}$ be the operator defined by (25). Suppose $\mathscr{F}$ and $\mathscr{D}_{(q,\lambda)}$ form an associated pair of operators, for each fixed $t \in [0, T]$, and the solutions of the corresponding equation $\mathscr{D}_{(q,\lambda)} u = 0$, satisfy an interior estimate of first order. Then the initial value problem (23)–(24) is solvable provided that the initial function is a meta-q-monogenic function. The solution exists in conical domains with the height sufficiently small.*

Note that the solution is obtained by construction, using the method of successive approximations, i.e., the solution is the limit of the sequence $\{u_n\}_{n=0}^\infty$ defined by:

$$u_0(t, x) = \varphi(x) + \int_0^t C(\tau, x) d\tau \tag{29}$$

and

$$u_{n+1}(t, x) = u_0(t, x) + \int_0^t (\mathscr{F}' u_n)(\tau, x) d\tau, \tag{30}$$

where $\mathscr{F}' u = \mathscr{F} u - C(t, x)$.

5 Concluding Remarks

In this work, we have given necessary and sufficient conditions on the coefficients of the operator $\mathscr{F}$ defined by (25) under which $\mathscr{F}$ is associated with the operator $\mathscr{D}_{(q,\lambda)}$. It means that $\mathscr{F}$ transforms meta-q-monogenic functions into meta-q-monogenic functions, for a fixedly chosen t.

Using the equation $\mathscr{D}_{(q,\lambda)}u = 0$ some derivatives could have been discarded from $\mathscr{D}_{(q,\lambda)}(\mathscr{F}u)$, and the sufficient conditions for $\mathscr{D}_{(q,\lambda)}(\mathscr{F}u) = 0$ could have been obtained by comparison of the coefficients. On the other side, by substituting special meta-q-monogenic functions, we showed that these conditions are also necessary. Therefore, we have found all linear first order operators of the form (25), which are associated to $\mathscr{D}_{(q,\lambda)}$.

Theorem 8.3 implies that each initial value problem (23)–(24) is solvable provided that the initial function is a meta-q-monogenic function, i.e. if it belongs to an associated space to $\mathscr{F}$. However, an example by Lewy [8] shows that there are differential equations as (23) not having any solution, in particular no initial value problem is solvable for these equations. It means that no associated space with an interior estimate for its elements, can exist.

The technique of associated spaces allowed us to solve the initial value problem (23)–(24) and the solution (the desired fixed point) is the limit of successive approximations defined by the Eqs. (29) and (30).

References

1. Abbas, U.Y., Yüksel, U.: Necessary and sufficient conditions for first order differential operators to be associated with a disturbed dirac operator in quaternionic analysis. Adv. Appl. Clifford Algebr. **25**, 1–12 (2015)
2. Blaya, R.A., Reyes, J.B., Adán, A.G., Kähler, U.: Symmetries and associated pairs in quaternionic analysis. In: Bernstein, S., Kähler, U., Sabadini, I., Sommen, F. (eds.) Hypercomplex Analysis: New Perspectives and Applications Trends in Mathematics, pp. 1–18 (2014)
3. Ariza, E., Di Teodoro, A., Vanegas, J.: First order differential operators associated to the space of $q-$ monogenic function. Adv. Appl. Clifford Algebr. **27**(1), 128–147 (2017)
4. Ariza, E., Vanegas, J., Vargas, F.: First order differential operators associated to the space of monogenic functions in parameter-depending Clifford algebras. Adv. Appl. Clifford Algebr. **26**(1), 13–29 (2016)
5. Bolívar, Y., Vanegas, C.J.: Initial value problems in Clifford-type analysis. Complex Var. Elliptic Equ. **58**, 557–569 (2013)
6. Bolívar, Y., Lezama, L., Mármol, L., Vanegas, J.: Associated spaces in Clifford analysis. Adv. Appl. Clifford Algebr. **25**(3), 539–551 (2015)
7. Brackx, F., Delanghe, R., Sommen, F.: Clifford Analysis. Pitman Research Notes in Mathematics, vol. 76. Boston (1982)
8. Lewy, H.: An example of a smooth linear partial differential equation without solution. Ann. Math. **66**, 155–158 (1957)
9. Nagumo, M.: Über das Anfangswertproblem Partieller Differentialgleichungen. Jpn. J. Math. **18**, 41–47 (1941)

10. Hung, N.Q., Luong, N.C.: First order differential operators associated to the Dirac Operator of quaternionic analysis. In: Proceedings of ICAM Hanoi 2004 Methods of Complex and Clifford Analysis, pp. 369–378. SAS International Publications (2006)
11. Son, L.H., Van, N.T.: Differential associated operators in clifford analysis and their applications. In: Proceedings of the 15th ICFIDCAA, Osaka, OCAMI Studies, Vol. 2, pp. 325–332 (2008)
12. Son, L.H., Tutschke, W.: Complex methods in higher dimensions- recent trends for solving boundary value and initial value problems. Complex Var. Theory Appl. **50**(7–11), 673–679 (2005)
13. Tutschke, W.: Solution of Initial Value Problems in Classes of Generalized Analytic Functions, vol. 4. Springer, Berlin (1989)
14. Tutschke, W.: Interior estimates in the theory of partial differential equations and their applications to initial value problems. Mem. Diff. Equ. Math. Phys. **12**, 204–209 (1997)
15. Tutschke, W.: Associated Spaces—A New Tool of Real and Complex Analysis. National University Publishers, Hanoi (2008)
16. Yüksel, U.: Solution of initial value problems with monogenic functions in banach spaces with L_p -Norm. Adv. Appl. Clifford Algebr. **20**, 201–209 (2010)
17. Van, N.T.: First order differential operators transforming regular functions of quaternionic analysis into themselves. In: Proceedings of ICAM Hanoi 2004 Methods of Complex and Clifford Analysis, pp. 363–367. SAS International Publications (2006)

Correction to: Clifford Analysis and Related Topics

Paula Cerejeiras, Craig A. Nolder, John Ryan
and Carmen Judith Vanegas Espinoza

Correction to:
P. Cerejeiras et al. (eds.), *Clifford Analysis and Related Topics*,
Springer Proceedings in Mathematics & Statistics 260,
https://doi.org/10.1007/978-3-030-00049-3

In an earlier version of these proceedings, the affiliation of the volume editor "Carmen Judith Vanegas Espinoza" was incorrect. This has been corrected. Correct Affiliation: Department of Mathematics and Statistics, ICB, Technical University of Manabí, Portoviejo, Ecuador. The erratum book has been updated with the changes.

The updated version of the book can be found at
https://doi.org/10.1007/978-3-030-00049-3

© Springer Nature Switzerland AG 2018
P. Cerejeiras et al. (eds.), *Clifford Analysis and Related Topics*,
Springer Proceedings in Mathematics & Statistics 260,
https://doi.org/10.1007/978-3-030-00049-3_9

Printed in the United States
By Bookmasters

Printed in the United States
By Bookmasters